GRC PRODUCTION & USES

Graham True

Routledge
Taylor & Francis Group
LONDON AND NEW YORK

Author

Graham True was introduced to concrete technology while working at the Central Electricity Research Laboratories in Leatherhead, Surrey. His involvement included project work and experimentation to determine the performance of concrete used to form prestressed pressure vessels and radiation containment vessels for nuclear power plant operated by the C.E.G.B.

A more specific interest in glassfibre reinforced cement began with the author's employment by E H Bradley & Sons, Swindon. In 1973 they started investigating the properties and potential of this material in order to produce a successful range of products.

Similar investigations were initiated by Mr True's next employer Thyssen (GB) Ltd. Specialization in grc permanent formwork became the main stay of the operation bringing out all the product's potential.

This work was continued and expanded to include design and detailing aspects of grc permanent formwork while working for Velmac UK, a member of Tarmac Industrial Holdings Ltd.

At present Graham True is employed as Materials Engineer in the Central Engineering Division of Tarmac Construction Holdings.

First published 1986
By Viewpoint Publications.
Reprinted 2003
By Routledge
11 New Fetter Lane, London EC4P 4EE

Transferred to Digital Printing 2003

13.023

ISBN 0 86310 004 X

Front cover photograph courtesy of Velmac (UK) Ltd.

Back cover photographs courtesy of Forton BV (two top illustrations), Thyssen (GB) Ltd (base left) and Velmac (UK) Ltd with Tarmac Industrial Holdings Ltd (base right).

Printed and bound by Antony Rowe Ltd, Eastbourne

Acknowledgements

The Author would like to thank the following for permission to use their illustrations or tables in this publication.

3.1 Mono Pumps Ltd, Arnfield Works, Audenshaw, Manchester M34 5JA

3.2 3.3 3.13 3.14 3.18 3.22 3.23 3.26 K & C Mouldings (England) Ltd, Spa House, Shellanger, Diss, Norfolk IP22 2DF

3.4 3.5 3.10 3.19 Power-Sprays Ltd, Thornbury, nr Bristol BS12 1NP

3.6 3.7 3.8 3.20 3.24 3.25 Putzmeister Ltd, Carrwood Road, Chesterfield Trading Estate, Sheepbridge, Chesterfield, Derbyshire S41 9QB

3.9 3.21 Drostholm Products A/S, DK–2950 Vedbaek, Denmark

3.15 Desoutter Ltd, 319 Edgeware Road, Colindale, London NW9 5ND

3.28 Procem Co. Ltd, 2 London Road, Hazel Grove, Stockport, Cheshire

3.29 Stelmo Ltd, Charing, Ashford, Kent TN27 0EJ

3.30 3.31 Marine Engineering Co. (Stockport) Ltd, Bulkeley Street, Edgeley, Stockport, Cheshire SK3 9AJ

3.33 TAC Construction Materials Ltd, PO Box 22, Trafford Park, Manchester M17 1RU

3.34 ARC Ltd, Head Office, Quarry Road, Chipping Sodbury, Bristol BS17 6AY

3.35 3.36 Banbury Buildings Ltd, Ironstone Works, Adderbury, Banbury OX17 3NS

3.37 Gusmer, c/o Baxenden Chemical Co. Ltd, Paragon Works, Bayenden, Accrington, Lancs BB5 2SL

5.1 Fibreglass Ltd, Prescot, St Helens, Merseyside WA2 0TT

5.4 Microfine Minerals & Chemicals Ltd, Mica Works, Raynesway, Derby DE2 7BE

6.1 Liner Croker Ltd, Runnings Road, Cheltenham, Gloucs GL51 9NR

6.2 6.3 6.6 Joshua Greaves & Sons Ltd, PO Box 2, Atlas Engineering Works, Ramsbottom, Bury, Lancs BL10 9BA

6.4 6.5 Tumac Maskin AB, Tjadervagen 35, 141 72 Huddinge, Sweden

7.1 7.2 7.3 From the CERA Translation No. 12. *Glassfibre Reinforced Cement* by K L Yu, L and D L Biryukovich. Published by Civil Engineering Research Association

9.1 9.2 9.3 9.4 9.5 9.6 9.7 9.8 11.9 13.27 13.28 13.29(i) 13.29(ii) 13.29(iii) 13.33 13.34 13.35 13.40 13.41 13.42 Thyssen (GB) Ltd, Bynea, Llanelli, Dyfed SA14 9SU

10.1 10.2 10.3(a) 10.3(b) 10.4 Michael Fordyce, Building Design Partnership, Vernon Street, Moor Lane, Preston PR1 3PQ

11.11 11.12 Building Research Station, Garston, Watford WD2 7JR

11.19 Monsanto, Dorcan Way, Swindon, Wilts SN3 5HN

11.20 Nene Instruments Ltd, 21–23 Park Road, Wellingborough, Northants NN8 4PW

11.21 J J Lloyd Instruments Ltd, Brook Avenue, Warsash, Southampton SO3 6HP

13.1 13.2 13.3 13.4 13.5 13.6 13.7 13.8 13.39 Glass Reinforced Concrete (GRC) Ltd, Wincham Lane, Wincham Northwich, Cheshire CW9 6DE

13.9–13.22 13.36 13.37 13.38 Owens Corning Fiberglas Corporation, Fiberglas Tower, Toledo, Ohio 43659, USA

13.24 13.25 13.26 Veldhoen-Isolatie BV, Raalte, Holland

13.44 13.45 13.46 15.14 Table 15.3 Forton BV, Paardestraat 31, 6131 HB Sittard, Holland

15.1–15.13 Tables 15.1 & 15.2 Taken from reports published by the Building Research Station, Garston, Watford WD2 7JR and Fibreglass Ltd, Pilkington Reinforcements Ltd, St Helens, Merseyside WA10 3TR

Contents

Preface

All new materials require a period of careful investigation in order to establish properties, product potential, design and manufacturing methods. On this basis market recognition and acceptance is achieved. Glassfibre reinforced cement has passed through this phase and is now treated as a material with significant inherent possibilities. Its potential has given rise to particular grc products such as cladding panels, asbestos replacement products, pipes and permanent formwork. The latter employs the properties of this material to such a degree that it warrants its placement in a special category — unequalled by current alternatives. Codes of Practice and British Standards are now including glassfibre reinforced cement and thereby giving full recognition to the material in its own right.

This book sets out the production of grc, exploring the various methods currently available. Basic 'run of the mill' together with some of the more specialized processes are included and the equipment available is described. Naturally, with a new material, innovation is continually bringing about new manufacturing processes, some of which are closely guarded by companies developing them into commercial propositions. Chapters are devoted to quality control, design, costing, curing and finishes. In addition, the final uses of the product are described focusing on both the general and the specific. This enables the reader to appreciate the possibilities of glassfibre reinforced cement to the full.

Glassfibre reinforced cement cannot be aligned with concrete or glassfibre plastic as so often is the case. It must be classed separately as a material with its own distinct properties. The author's intention, through this book, is to encourage further acceptance of this novel and useful material.

Graham True
1986

1

Background

In 1967, Dr A J Majumdar of the UK Building Research Establishment (BRE), Watford, began investigating the effect of zirconia additions to glass formulations. The overall aim was to achieve a glassfibre that had sufficient alkali resistance to withstand the environment found in hydrating Portland cement. Some of these formulations were successfully fiberized and their potential use for reinforcing Portland cement was demonstrated. The National Research and Development Corporation (N.R.D.C.) filed patent applications to cover these achievements.

Development work encouraged trials to determine whether the formulations could be produced economically on a commercial scale. BRE and N.R.D.C. set out, with Pilkington Brothers Ltd, to undertake further commercial exploration. As a result, an agreement was reached whereby Pilkington Brothers Ltd, through their subsidiary Fibreglass Ltd, was licensed exclusively to develop world-wide the commercial potential of this alkali resistant fibre.

By 1971 Pilkington Brothers Ltd had achieved:

1. The development of a viable, commercially fiberizable alkali resistant glassfibre, with the trade name of Cem-FIL

2. The furthering of the test programme set up by BRE to assess the long-term durability of composites containing Cem-FIL, when exposed to a range of environmental conditions

3. The development of further prototype products and processes employing Cem-FIL alkali resistant fibre.

The third stage followed on from earlier work carried out at BRE using non-alkali resistant 'E' glassfibres. While the alkali resistant fibre formulation was under development, production techniques were investigated using the non-alkali resistant 'E' glassfibre in ordinary Portland cement (oPc) and gypsum plaster. This search for a suitably resistant glassfibre for incorporation into oPc proceeded from earlier tests in the UK where 'E' glassfibre had been incorporated into high alumina cement mortar.

Research in Russia had also been in progress since 1941 when glassfibre rovings were used in an attempt to reinforce concrete. The experiment had, however, met with little success. Glassfibre rovings were therefore replaced by glassfibre reinforced plastic rods, but the presence of plastics caused very considerable creep. Finally, research culminated in a publication in 1965 which detailed the development of 'dispersed-stress' reinforcement and, consequently, a material with a completely new structure and completely new properties. This material, now referred to as gfr cement, has glassfibres dispersed over its entire section and orientated mainly along the line of action of the tensile forces. The resultant material is one in which the densely laid fibres reinforce thin layers of hardened cement paste.

Principal types of binder used to make gfr cement were high alumina cement, gypsum-aluminous cement and waterproof expanding cements. The Russian report notes that Portland cements lead to failure of the glassfibre reinforcement owing to the aggressive action on the glassfibre. It also goes on to state that protective films or high-alkali fibres exceeding 25 microns in diameter may be used with Portland cement.

Overall Russian research probably prompted many investigators to study the possibilities of grc and from this evolved the alkali resistant fibres mentioned earlier.

Pilkington Brothers Ltd have now produced a new fibre named Cem-FIL 2 which will replace Cem-FIL. This fibre is based on the original Cem-FIL but with the added benefit of a special surface coating that prolongs the ductile failure pattern previously achieved. Other companies, in particular Forton in Holland, have adopted the technique mentioned in the Russian Report – not only coating the surface of an 'E' glassfibre with a suitable polymer but additionally dispersing a polymer solution throughout the matrix. The extra cost of the polymer dispersion is offset by the lower cost of the 'E' glassfibre to achieve a similar overall cost for the composite at the 5% additional level.

Various incorporation techniques have been developed ranging from pre-mixing the fibres into the matrix to hand laying and rolling into the cement paste, from winding and pressing using nets and continuous aligned rovings to spraying simultaneously chopped fibres and cement mortar causing mixing in mid-stream before depositing onto the mould surface. The latter method of spraying has gained enormous popularity due to the ease with which it can be performed, the availability of production equipment from the glassfibre reinforced plastics (grp) industry and the superior properties that can be achieved by this method of production. Early work at BRE was based on this spray-up technique but was further assisted by vacuum dewatering of the composite in order to reduce the water content, increase the density and improve the physical bond between the fibre and the composite.

2

Requirements for production

When planning a factory lay-out, the inter-relationship of the various production stages and their individual function and scope must be clearly defined. In the following notes, which are intended to assist this preparation, the functions of a grc production facility are divided into three groups:

Administration — Production — Storage

These three groups have been further sub-divided:

Administration:	Office administration (costing, buying, marketing, management, design)
	Quality control testing and recording
Production:	Spraying and mixing
	Production quality control laboratory functions
	Mould manufacture
	Mould cleaning and oiling
	Overnight storage, demoulding and making good
	Curing
Storage:	Raw materials storage
	Mould storage
	Stockyard operations and despatch

2.1 ADMINISTRATION

This section highlights aspects particularly relevant in grc production. It is not felt necessary to cover 'pure' administration requirements which have a general application.

2.1.1 Office administration

The manufacturing unit should be closely associated with administrative personnel in order to produce a unified team that benefits from feed-back concerning shop-floor operations. This encourages a speedy eradication of production problems and ensures that errors and misjudgements at the design and planning stages of a contract do not carry forward into subsequent contracts.

2.1.2 Quality control testing and recording

This aspect of the quality assurance function is categorized under administration because its purpose is defined as:

1. To test grc test coupons for flexural strength, density and so on, at 7 and 28 days as specified

2. To analyze the results recorded from the above and keep a running account (graphical or otherwise) of the quality of the output. Thus an overall 'picture' concerning the standard reached is obtainable and this assists the design department in specifying, with confidence, new production requirements.

If a full statistical analysis of the type shown in *Chapter 11* is maintained then a good understanding of production capabilities and the cost/effective use of fibre and other raw materials can be seen.

This department should also administer the functions of the production quality control laboratory.

2.2 PRODUCTION

Any discussion on this topic is directly related to the method or methods being used and varies accordingly. For this reason, only notes of a general kind are applicable.

Figure 2.1 shows the inter-relationship of typical production steps and these apply to most production set-ups including hand-spray, automated spray and specialized procedures. Consequently this diagram should also be seen as a guide to the physical lay-out of a production process. When considering every new grc production set-up a similar plan should be worked out taking account of the specific manufacturing process to be employed.

2.2.1 Spraying and mixing

Since almost all activities concerned with production relate to spraying and mixing, this particular stage should be considered as the focal point of the production process. A particular manufacturing facility may require a number of spraying and mixing set-ups in order to process concurrently a range of products. This could enable other stages in the production process to be shared but, nevertheless, the following aspects have to be considered in order to plan out the physical location of the equipment and the area required:

1. The type of spray rig to be used — either a simple hand-spray facility or a compact spraying/mixing set-up (refer to *Chapter 3, Figure 3.27)*

2. The area required to position and carry out mould spraying. Turntables, trestles and conveyors may be required

3. A location at hand for storing raw materials needed in the mixing and spraying operation and taken from the main storage area

4. Weighing scales, if batching is to be carried out alongside the mixer set-up (pre-bagged, pre-weighed quantities may be more effective)

5. Washing-down facilities including drainage and settlement pits

6. Easy access to and from the area for moving moulds before and after filling, as well as supplying raw materials. Consideration should be given to the best method of transporting mould and materials, whether by conveyors, fork-lift trucks, monorails or hand barrows.

From the above it becomes apparent that the mixing/spraying area is one of transition and requires careful planning in order to establish an efficient flow to the process.

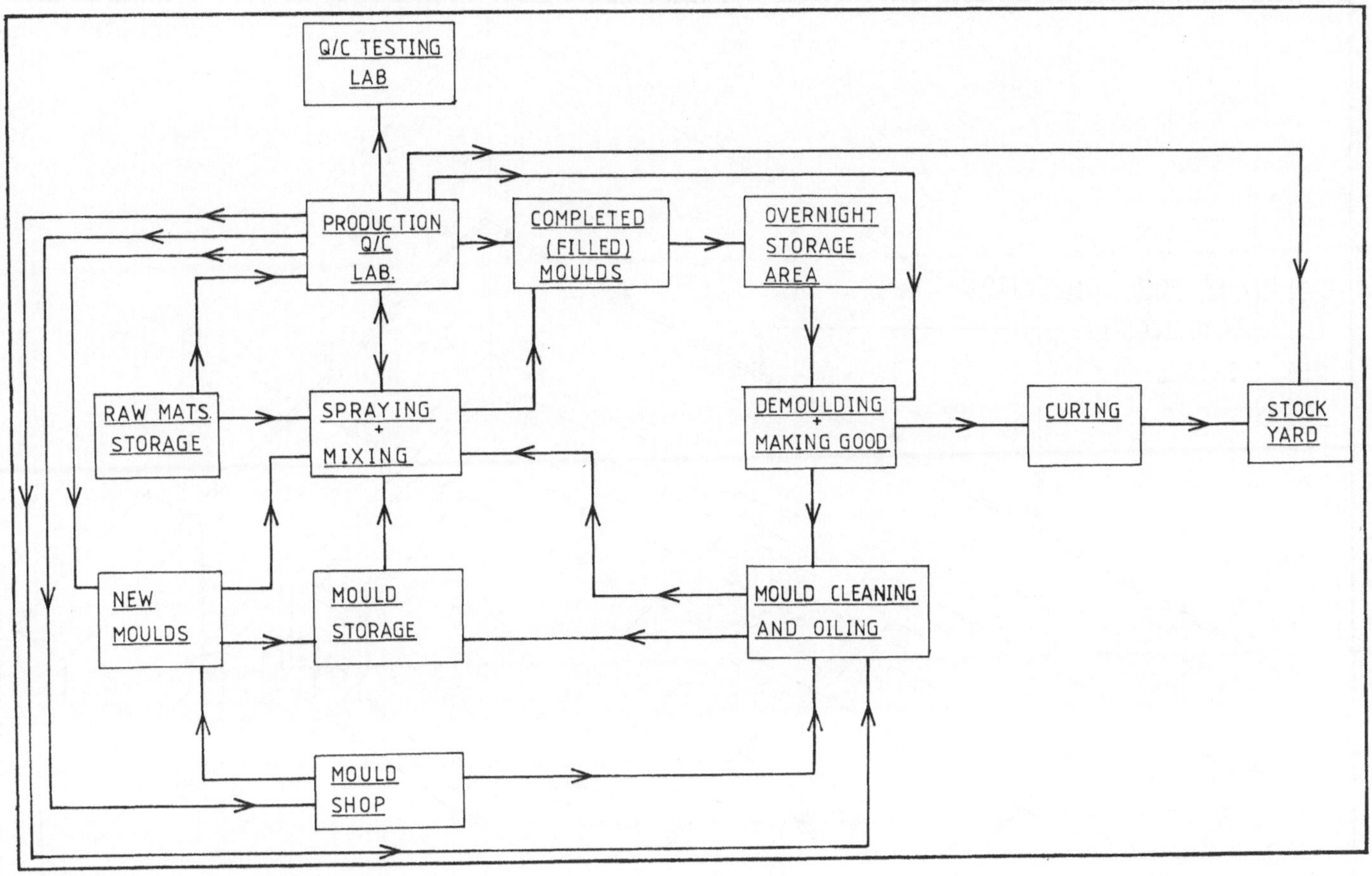

Figure 2.1 Inter-relationship of production steps.

2.2.2 Production quality control laboratory functions
When adapting Figure 2.1 for the actual process required it is necessary to retain the production quality control laboratory within the works area. This is important because it has a supervisory capacity that relates to practically all functions in the factory.

Facilities required are balances, drying ovens, washing equipment for washout tests and a filing system to house the records of checks made and specifications that are to be monitored (for example, mould sizes, tolerances and mix design data). This laboratory will process the test board coupons prior to curing in the testing laboratory and future strength testing.

2.2.3 Mould manufacture
This unit would obviously be responsible for the manufacture and alteration of the production moulds and should be suitably equipped with the necessary tools to process wood, steel, plastics, grp or whatever materials are to be employed. The unit can be conveniently located away from the manufacturing area and, quite possibly, in a separate building.

2.2.4 Mould cleaning and oiling
Whether for new moulds awaiting filling or old moulds already in the manufacturing cycle, a location for cleaning and oiling is required prior to spraying. This should be sited adjacent to the spraying area to ensure a continuity of production.

2.2.5 Overnight storage, demoulding and making good
After filling, moulds require storage overnight prior to demoulding. Ideally this should be next to the spraying area enabling the moulds to be cleaned rapidly and oiled, and returned for further spraying.

It must be emphasized that unless good substantial steel or plastic moulds are used, checks should be made before sending the moulds back into production to ensure dimensional accuracy is maintained. Any making good of the demoulded products is also best carried out as soon after demoulding as possible and certainly before despatch into the curing chamber.

2.2.6 Curing
Again, this is a function that can be located away from the main production area. It is a stage in the manufacturing process that requires careful planning and consideration since it needs:

1. Flexibility — to cope with a range of sizes and quantities of products
2. Consistency — to ensure that the conditions of suitable curing are maintained (i.e., 100% RH at 20°C)
3. Cost/Effectiveness — to provide the most practical solution that requires little maintenance and has low running costs
4. Compartmentation — to allow each day's output to be cured separately and independently avoiding confusion.

Figure 2.2 shows a suggested curing chamber layout suitable for grc 7-day curing. The area can be divided by rubber curtains to produce the required number of booths, and the front should be sealed with insulated durable doors. The plumbing requirements are given in Figure 2.3. When using flexible hoses within the curing chamber to direct the spray of water, it is possible to adapt the conditions to suit all types and sizes of production units.

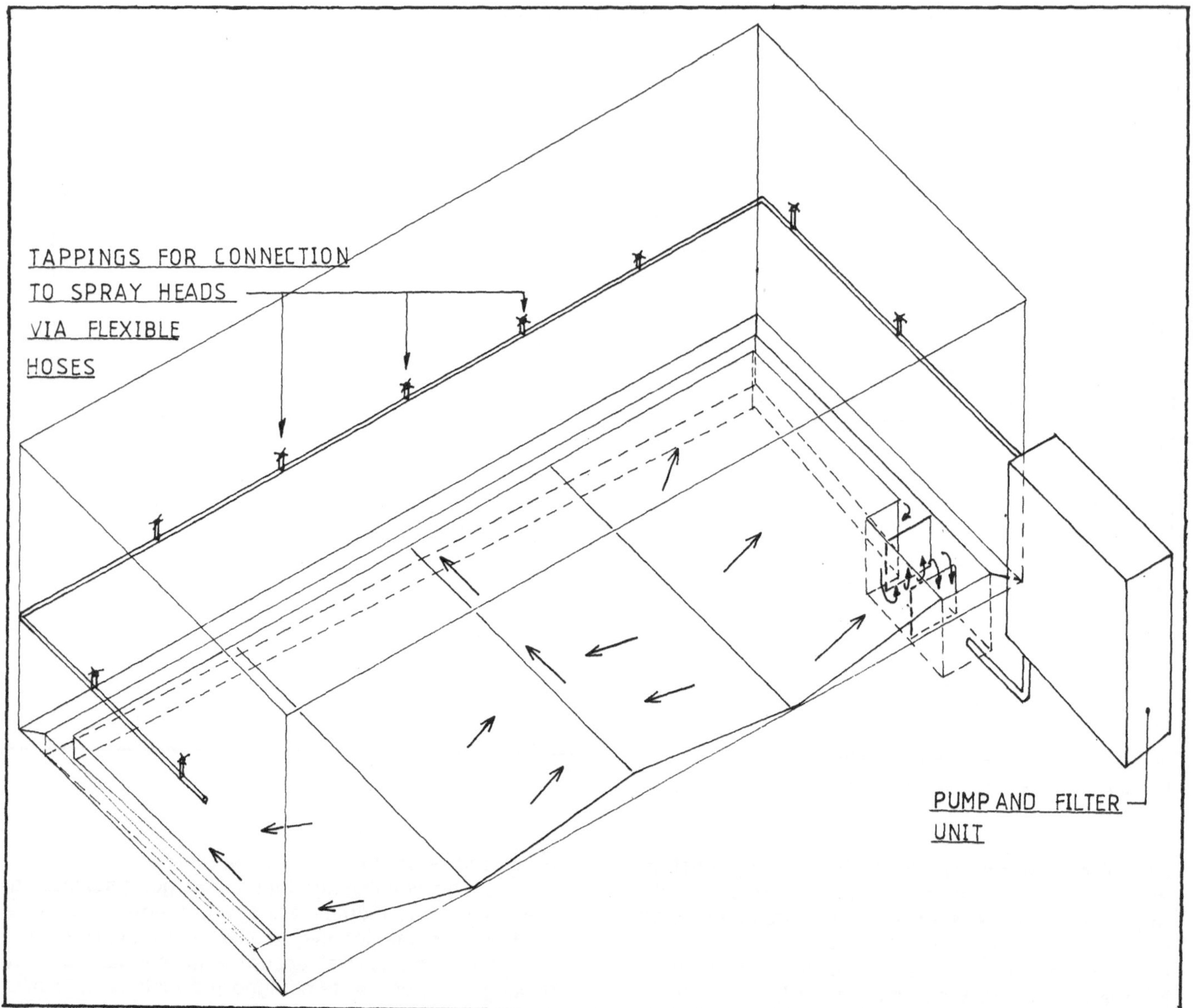

Figure 2.2 Lay-out of curing chamber.

2.3 STORAGE

Important aspects to be considered are:

- Raw materials storage
- Mould storage
- Stockyard operations and despatch.

2.3.1 Raw materials storage

Clearly a dry area is necessary for this and, if the scale of operation is sufficiently large, it is worth considering bulk silo storage of sand and cement. However, it must be noted that a precise and sensible method of extraction from the silos is required. This enables batch quantities to be drawn and placed into the mixer. It will, however, mean a large capital outlay.

2.3.2 Mould storage

Ideally moulds should be stored away from the production area. Yard storage may suffice, but protection in the form of covering and pretreatment (i.e., oiling of steel moulds) is vital. Each one must be clearly labelled to provide a register of stock moulds. Moulds can often be adapted for future needs, unless improper storage renders them useless.

2.3.3 Stockyard operations and despatch

This function should be under the control of one supervisor. Naturally sensible stocking of products in the yard, correctly labelled and in a fit state for despatch, is important and a suitable site should be chosen away from delivery or heavily trafficked areas. It simply requires careful planning of the available space location, with racking and stillages used to enable quick despatching.

Figure 2.3 Curing room plumbing lay-out.

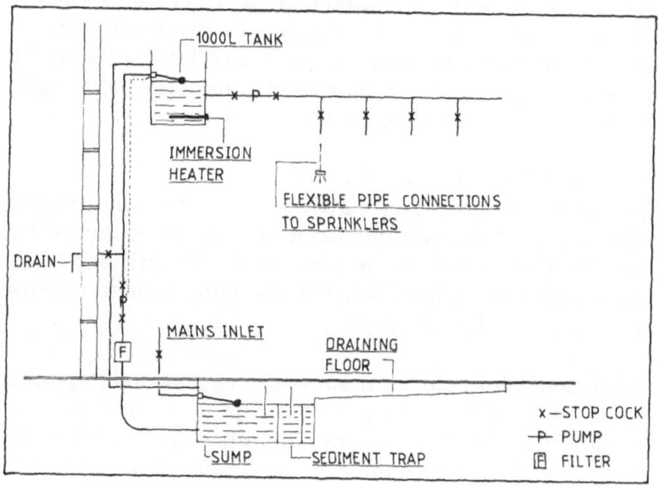

3

Production equipment

In this Chapter it is intended to give a fairly broad guide to the equipment available for the production of grc. Since hand-spray work and 'mechanized' hand-spray work are at present the major volume production processes employed, most consideration is given to the equipment they require. Sections are included, however, to illustrate specially developed processes that are either now being brought into use or have at least been described as viable manufacturing techniques for the production of grc components.

3.1 PUMP UNITS

The pump unit is the piece of equipment that processes the matrix or mortar component of the grc composite. Freshly mixed mortar, or in some cases the raw materials, is transported by the pump unit at discrete set rates to the spray head or gun. The unit can therefore be regulated to produce a precisely maintained mortar output at the gun nozzle.

There are few machines available at present but nearly all employ the worm action pump, and the Mono Pump MT 420 wide-throat pump is common in UK manufactured units.

The *MT 420 Pump* is designed for mixes with difficult flow properties. It is essential that the speed of the pump is carefully selected to ensure the delivery rate conforms with the relative velocity at which mortar can pass through the pipeline. Very stiff mixes are directly influenced in their rate of output by the pipeline friction with only marginal relationship to static head. Consequently, when low water/cement ratio mixes are processed, admixtures may be required to assist the pumpability of the mix.

The components of a Mono Pump MT 420 are shown in Figure 3.1. The wide throat entry allows mortar to fall easily onto the auger-conveyor which transports it to the rotor/stator. The Mono pumping element consists essentially of an elastomeric stator (35) in the form of a double internal helix and a single helical rotor. The stator is a single component clamped to the pump and end cover by long bolts fitted with hand nuts for quick dismantling and therefore easy cleaning and maintenance. The rotor (39) maintains a constant seal across the stator and this seal travels continuously through the pump giving uniform positive displacement. The rotor is removed by disconnecting the rotor head from the connecting rod.

The pump must *never be run in dry condition as the stator will instantly be damaged.* It is crucial to ensure that the hopper is filled with mortar so that the auger-conveyor engages with the mortar to fill the pump immediately. A few turns of the rotor by hand will assist the initial lubrication of the stator until the pump primes itself. When the pump is stopped, sufficient mortar is trapped to provide lubrication on starting-up again. The pump is arranged for anti-clockwise rotation facing the driving end.

The drain plug may be replaced by a stop valve which, when connected to the water mains, provides a quick flushing-down at the end of the work shift. It also enables the stator to be fed with water at the beginning of the morning shift to prevent undue stator wear at start-up.

The ball bearings require periodic cleaning and it is advisable to establish the correct maintenance schedule by regular examination.

Mono Pumps Ltd are located at Arnfield Works, Audenshaw, Manchester M34 5JA.

3.1.1 Downland grc spray pump unit

The pump (Figure 3.2) is mounted on castor wheels and comprises a mortar hopper with sieve, safety guard and vibrator, feeding the mortar to a Mono Pump MT 420. The Mono Pump is driven by an enclosed fan cooled electric motor. To maintain the essential accuracy of slurry output, the pump is driven through a variable speed thyristor control linked to the drive motor. The thyristor control permits manual setting. An electronic speed indicator displays the rotor speed (Figure 3.3).

By thyristor speed control, transmission losses are minimized allowing the use of a 1.5 hp single phase 240 V motor. All the controls are grouped on one panel. They include air regulators and pressure gauges for each of the compressed air systems, electric switch gear for auto and hand control, pump speed regulator and speed indicator, and compressed air filter and lubricator. The control panel may also be supplied as a wall mounted unit, or as a separate console when it is necessary to group controls on automated plants.

The specification for this unit is given in Table 3.1.

The Downland grc spray pump unit is obtainable from K & C Mouldings (England) Ltd, Spa House, Shellanger, Diss, Norfolk IP22 2DF.

Table 3.1 Specification for the Downland grc spray pump unit.

Electric supply	240 V
Motor	1·5 hp
Main fuse	10 amp
Thyristor fuse	20 amp
Weight	325 kg
Size: Height	1067 mm
Size: Length	1220 mm
Size: Width	991 mm
Maximum mortar output with standard pulley arrangement	15 kg/min.

3.1.2 Power-Sprays grc spray pump

The Power-Sprays company has marketed grc spray units for a number of years using initially the narrow

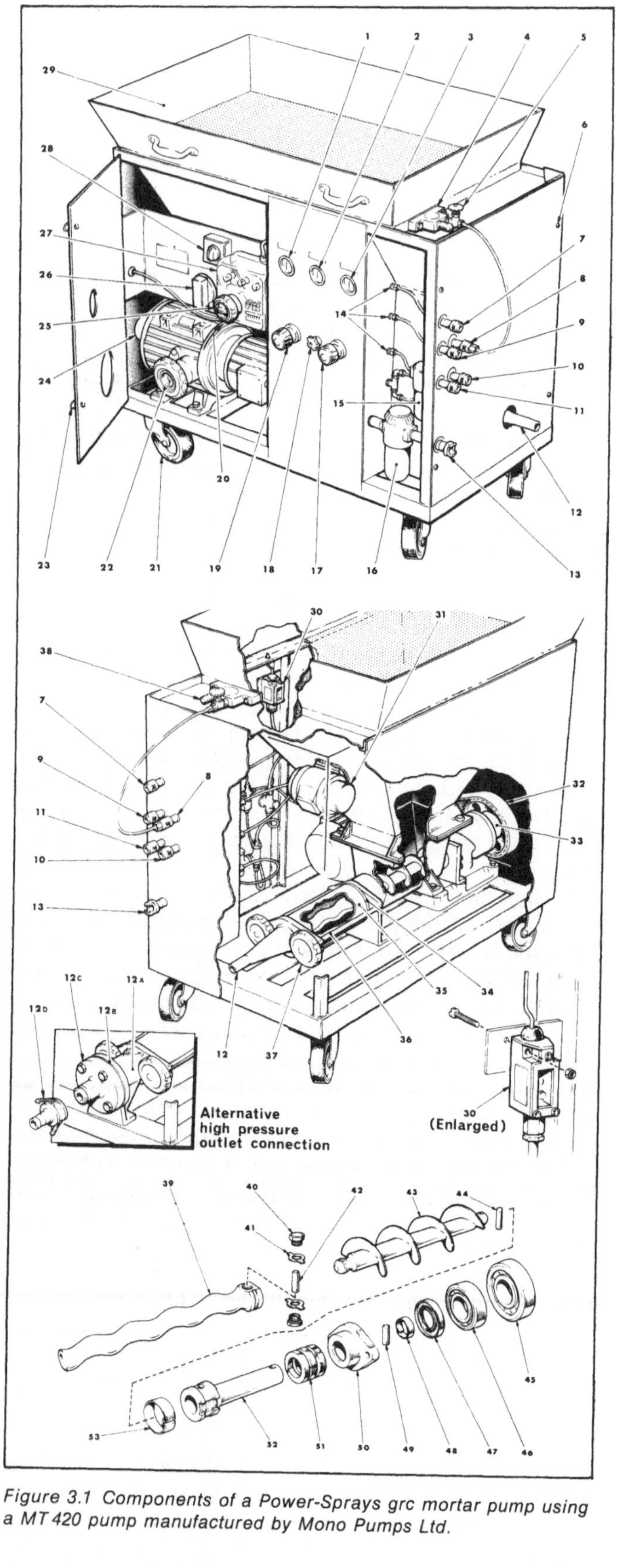

Item	Description	Qty
1	Chopper air pressure gauge	1
2	Spray air pressure gauge	1
3	Air mover air pressure gauge	1
4	Sieve vibrator	1
5	Sieve vibrator control valve	1
6	QR panel securing screws	—
7		
8		
9	QR air outlet couplings	5
10		
11		
12	Pump outlet connection/outlet flange	1
12A	Pump outlet flange	1
12B	Pump outlet connection gasket	1
12C	Pump outlet connection	1
12D	Output hose connector	1
13	Main air inlet claw coupling	1
14	Nylon tube terminations	—
15	Air line lubricator	1
16	Air filter	1
17	Air mover pressure regulating valve	1
18	Spray air pressure regulating valve	1
19	Chopper air pressure regulating valve	1
20	As for no. 25	—
21	Swivel castors	4
22	Speed control handwheel/indicator	1
23	QR door securing screws	2
24	Gearbox power output pulley	1
25	Main isolator switch	1
26	Air pressure switch	1
27	Electronics control box	1
28	Hopper vibrator control switch	1
29	Sieve	1
30	Sieve safety switch	1
31	Hopper vibrator motor	1
32	Toothed drive belt	1
33	Pump driven pulley	1
34	Stator support ring	1
35	Stator	1
36	Stator tie rod assembly	2
37	Stator tie rod handwheel	2
38	Sieve vibrator air exhaust silencer	1
39	Pump rotor	1
40	Rotor pin cap	2
41	Rotor pin cap washer	2
42	Rotor pin	1
43	Pump auger conveyor	1
44	Auger pin	1
45	Bearing (large)	1
46	Bearing (small)	1
47	Grease seal	1
48	Yoke pin retaining clip	1
49	Yoke pin	1
50	Gland	1
51	Gland packing	—
52	Drive yoke	1
53	Auger pin retaining clip	1

Figure 3.1 Components of a Power-Sprays grc mortar pump using a MT 420 pump manufactured by Mono Pumps Ltd.

throat Mono Pump MT 42 and changing to the MT 420 about 6 years ago. The unit is offered with a Carter gearbox and 3 hp electric motor. The control panel, in the case of the early models, was wall mounted but is now obtainable set on the pump chassis. Figure 3.4 illustrates a 4V04C model from 1975.

This company has manufactured a number of special pumps. Figure 3.5 shows a particular unit that enables the stator to be easily removed for cleaning and has the controls housed in the pump cabinet, protected and viewed through a wire mesh screen. Vibrating screens are now offered on top of the hopper feed.

Power-Sprays Ltd are based at Orchard Works, Whale Wharf, Littleton-on-Severn, Thornbury, Bristol BS12 1NP.

3.1.3 Pilkington grc pump unit

This unit has been developed by Pilkington Brothers to meet their design specification. It is intended for both hand-spray and mechanized grc production.

The unit comprises a feed hopper, worm-type pump, electric motor, variable speed gearbox, starter and pneumatic system for operation of cement spray and glassfibre chopper. A matching spray/chopper head is also available for hand-spray processing. The pump is

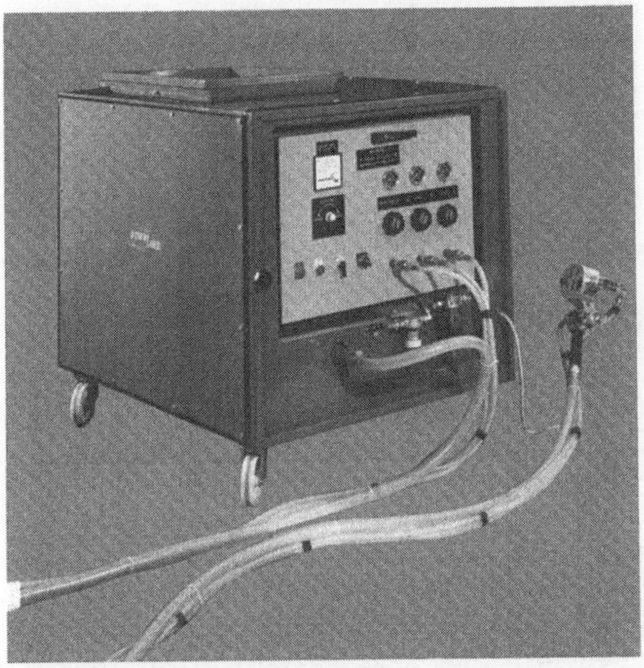

Figure 3.2 Downland grc spray pump unit.

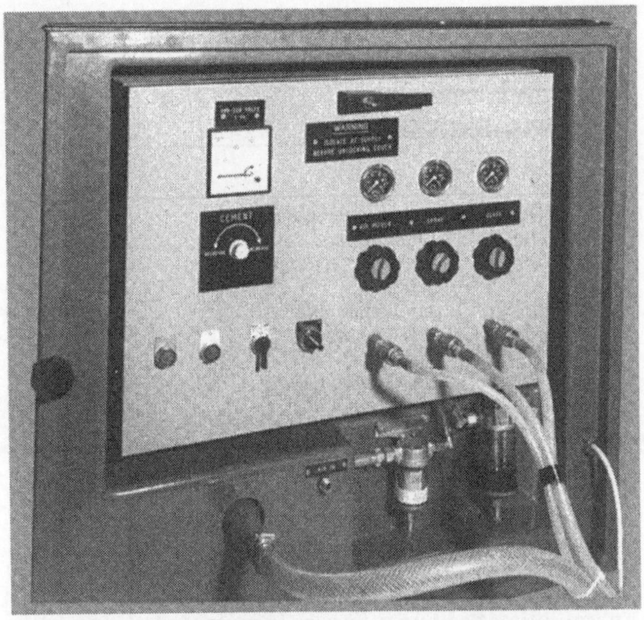

Figure 3.3 Control panel of the Downland grc spray pump unit.

Figure 3.4 MK 7 pump unit—early 1975.

Figure 3.5 A special Power-Sprays Ltd grc spray pump unit.

capable of handling, with ease, thick pastes of low water/cement ratio. The unit is totally enclosed for safety in use and for easy cleaning. Electric and pneumatic equipment is housed in separate compartments to aid maintenance. The pump unit can be easily demounted for cleaning. Its specification is given in Table 3.2.

This pump unit has been available from B R Fabrications Ltd and finally through Pilkington Brothers Ltd but is now no longer in production. However, enquiries can be directed to Cem-FIL Technical Service, Lathom Laboratories, Ormskirk, Lancs.

Table 3.2 Specification for the Pilkington grc pumping unit.

Power supply requirements	Standard model 440 V 50HZ 3-phase E model 380 V 50HZ 3-phase
Compressed air supply requirement	0.56–1.12 m^3/min. at 0.6–0.7 N/mm^2
Worm pump	Mono Pump MT 420
Electric motor	3 hp 1420 rpm
Gear box	Carter-Renolds F13–5.71 reduction box with continuously variable ratio and dial indicator
Transmission	Fenner timing belt
Hopper	Gravity feed to auger complete with mesh screen
Pneumatic equipment	Distribution system for supply to spray nozzle, glassfibre chopper and air-mover Individual control valves and pressure gauges; filter and lubricator in supply to glassfibre chopper Quick release couplings on all lines
Controls	Main switches mounted on unit; remote control from spray/chopper head
Overall dimensions	1170 mm × 970 mm × 1070 mm high
Weight	395 kg

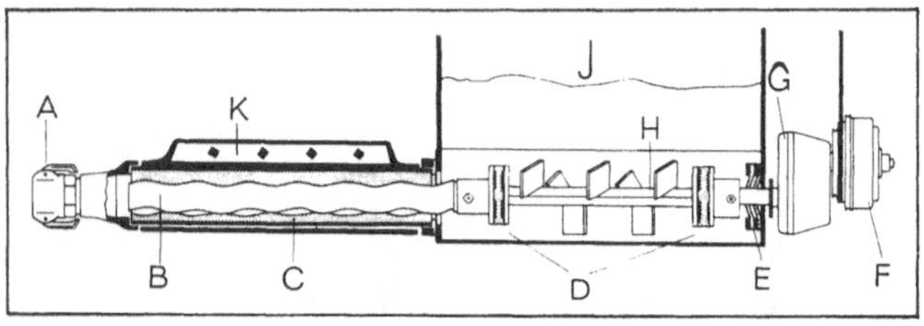

Figure 3.6 Putzmeister pump components.

Item	*Description*
A	*PM quick action coupling*
B	*Rotor*
C	*Adjustable stator*
D	*Flexible discs*
E	*Disc seal*
F	*Clutch with pulley*
G	*Gear box*
H	*Agitator*
J	*Hopper*
K	*Clamp*

3.1.4 Putzmeister grc pump units

Putzmeister offer two pump units suitable for grc manufacture:

- P 4 fibreglass concrete sprayer
- P 11 fibreglass concrete sprayer.

Both these units, although having distinct and different drive units, employ the same principle for pumping mortar (illustrated in Figure 3.6). The method used is basically the same as for the Mono Pump described earlier, but the Putzmeister has the advantage of using an adjustable stator, which accommodates wear and therefore has a longer service life.

In a factory environment the P 4 machine (Figure 3.7) is possibly more suitable but this does not preclude consideration of the P 11. The P 11 has, however, been successfully used to spray-apply premix grc (i.e., where the fibres are blended with the sand and cement at the mixing stage) and this has been carried out in situ at a number of locations. Figure 3.8 shows the P 11 machine. It can operate two pump units, or stator assemblies. This means two separate spray operations could be supplied from the one pump, an advantage most welcome in a factory production set-up.

Figure 3.7 P 4 fibreglass concrete sprayer.

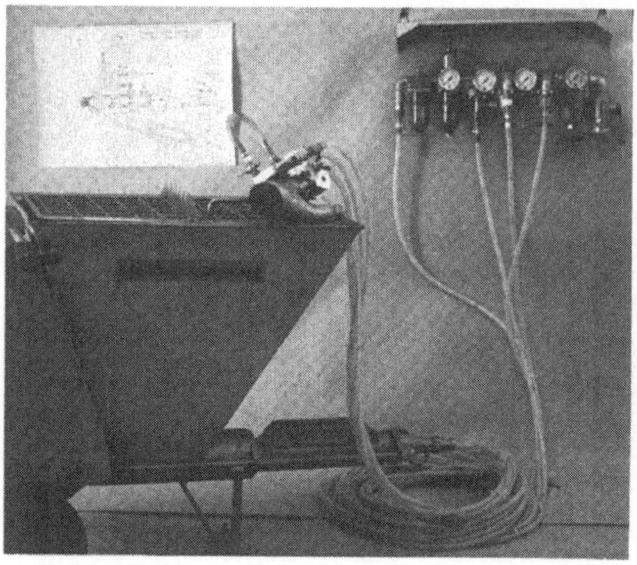

Figure 3.8 P 11 fibreglass concrete sprayer.

The output of the P 11 pump ranges up to 55 litres/min. compared with the P 4 pump which varies up to 12 litres/min. Naturally, both outputs are affected by the stiffness of the mix being pumped. The P 4 machine could benefit from a thyristor control to enable it to be 'trimmed' to suit the chopper gun output and therefore obtain exactly the required fibre content. Both machines can be operated from the spray head via air switches connected to the spray nozzle atomizer line. The standard P 4 requires 380 V, 3-phase supply whereas the P 11 is available with a 13.5 DIN hp diesel engine or a 5.5 kW 380 V motor.

Putzmeister equipment is obtainable from Putzmeister Ltd, Carrwood Road, Chesterfield Trading Estate, Sheepbridge, Chesterfield, Derbyshire S41 9QB.

3.1.5 Drostholm pump units

The Drostholm company has produced two pump units, the MG 4 and the MG 5. These units need to be fed with dry pre-blended sand and cement. Dry mortar is then passed via a special metering wheel into the built-in mixer. Mix water is gauged by a flowmeter which, together with the rapid mixing action, produces a homogeneous wet mortar.

Cleaning is rapid because the mixing chamber is the only unit requiring washing, leaving the dry material in the storage hopper ready for use when spraying is resumed. The pump can be controlled by the spray operator.

The specification for the MG 4 pump unit is given in Table 3.3 and the MG 5 pump in Table 3.4.

Table 3.3 Specification of the MG 4 pump unit*

Height	1420 mm
Length	1050 mm
Weight	190 kg
Working pressure	15–40 bars
Water input pressure	2 bars
Motor	4 kW
Voltage	as required
Fuses	3 × 25 A
Output	8–50 litres/min.

*See Figure 3.9

Table 3.4 Specification of the MG 5 pump unit.

Motor	3 kW
Voltage	as required
Output	8 litres/min. or according to requirements

The unit is normally controlled from the spray gun. It can be used for the mixing and pumping of mixtures with a particle size of max. 0.9 mm.

It is obtainable from Drostholm Products A/S, DK 2950 Vedbaek, Denmark.

Figure 3.9 MG 4 Drostholm grc pump unit.

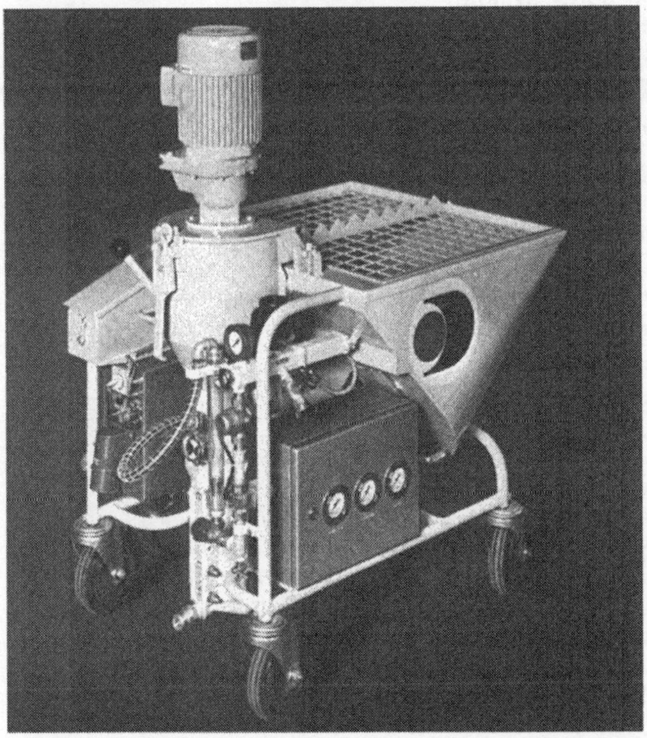

3.1.6 Power-Sprays mixer and pump unit

Power-Sprays have been developing a mixing pumping unit that receives unblended sand and cement which is stored in the machine in separate hoppers. The metering units within the machine draw off the sand and cement in the required pre-set ratio and feed it into a blending unit. The material is blended, and water added into the final stage of the machine via a flowmeter at a preset level. A high shear mixer incorporates the water into the mix to produce the required consistency enabling the pumping unit to transport the mortar to the spray head. The machine is a 'one man operation' controlled by the sprayer, thus enabling the whole process for the production of a grc mortar to be performed by one machine and one man (Figure 3.10).

It is supplied by Power-Sprays Ltd, Whale Wharf, Littleton-on-Severn, Bristol BS12 1NP.

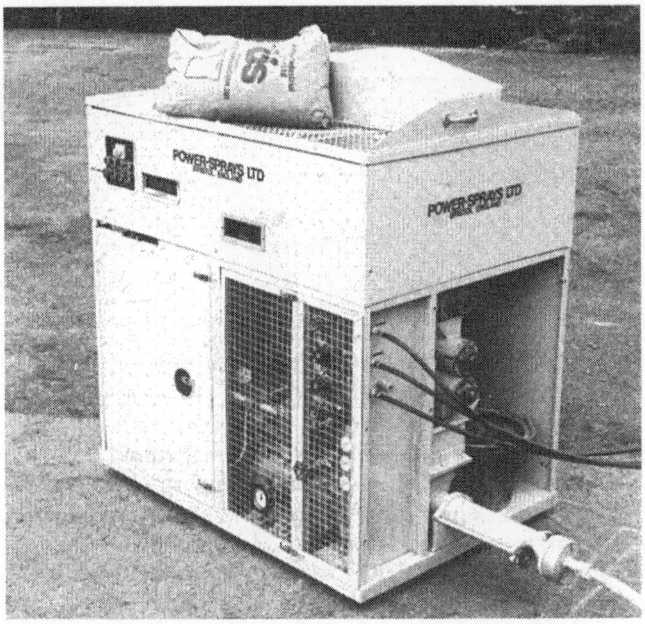

Figure 3.10 Power-Sprays mobile 'grc factory' pump unit.

Figure 3.11 Typical plastering machine.

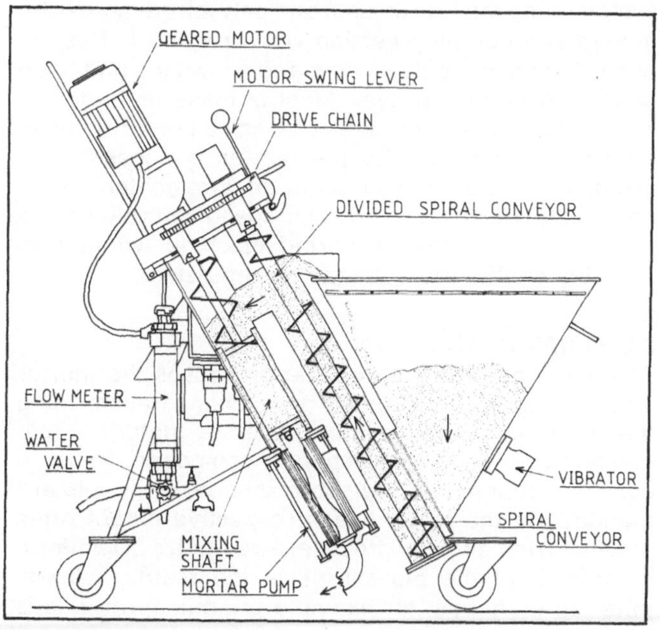

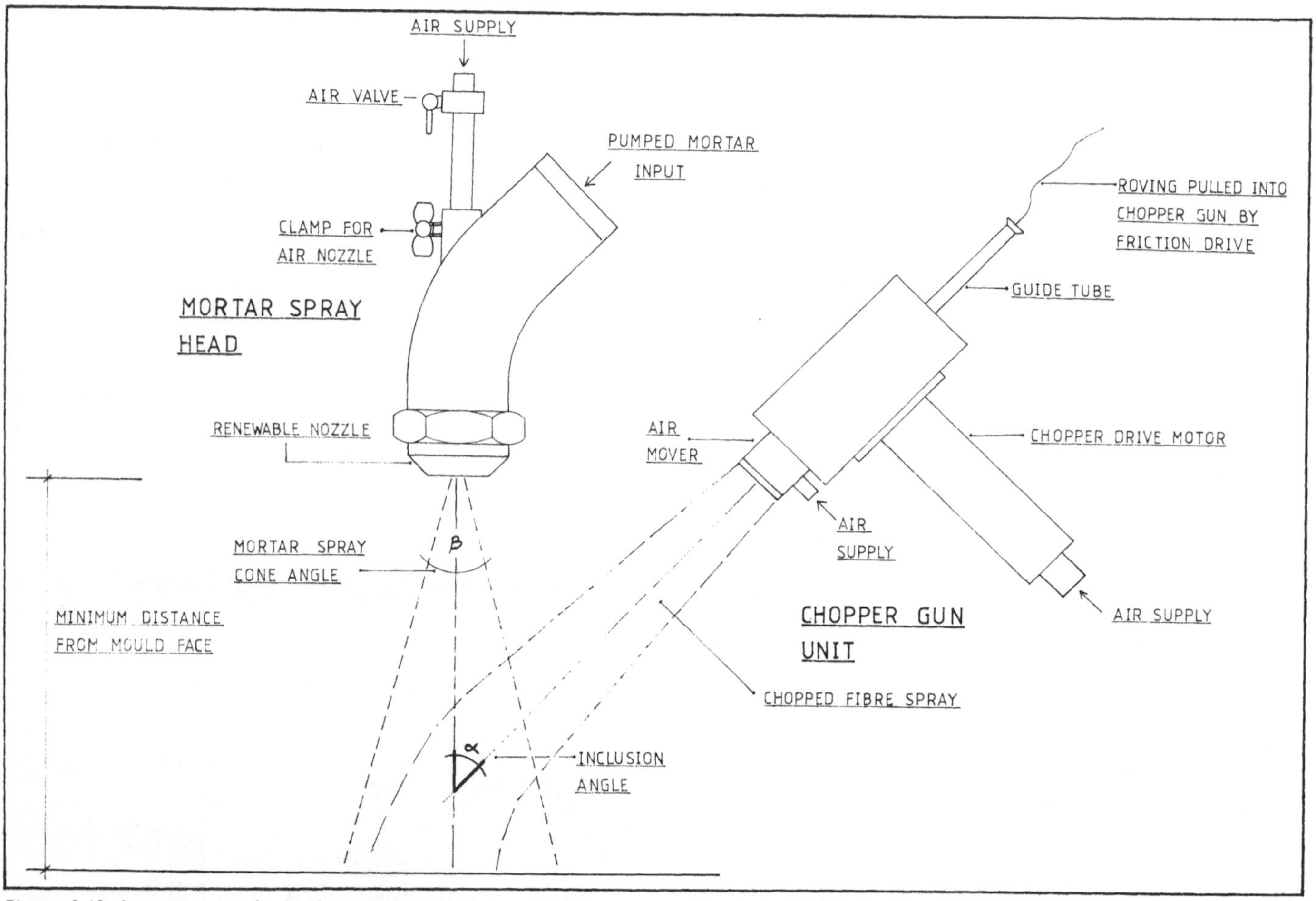

Figure 3.12 Arrangement of a basic grc spray head.

3.1.7 Plastering machines for grc production

The recently developed Power-Sprays mixer/pumping unit (just described) and the Drostholm MG 4 and MG 5 units can be seen as adaptions of the principle of plastering machines. There are numerous plastering machines available in Europe that could be adapted for grc production.

Figure 3.11 shows a typical unit. It illustrates the storage hopper and the screw feed which draws the dry mortar from the hopper to the mixer and pumping section. The principle employed means that dry mortar is cycled round the first column and only introduced to the mixing and pumping section when required. This reduces the cleaning down time and allows the unit to be controlled by the sprayer. Most of these units do not have a variable mortar output which can be adjusted to match that from the chopper gun as is necessary to produce and maintain the required fibre content in the composite. However, a thyristor speed controller fitted to the motor of the plastering machine would presumably enable the output to be regulated.

3.2 SPRAY HEADS

The spray head is the unit that processes the mortar and glassfibre (in spray-up production) transforming these two components from separate materials into one homogeneous well-dispersed composite. This process is a continuous flow, so that separate inputs are blended via the spray heads respective output rates and inclusion angle to produce, at lay-out, a 2D uniform dispersion of the glassfibres in the matrix. Clearly other spray heads, for example pre-mix spray heads and the unit used in pipe production, operate in a different way. These are described elsewhere in the text.

The components of a spray-up grc spray head are seen in Figure 3.12 and in summary are as follows:

(*a*) *Chopper gun unit* (which takes the roving, chops it into uniform lengths of fibre and sprays it into the mortar spray system)
It is sub-divided into:

- Chopping head
- Chopper drive motor
- Air mover.

(*b*) *Mortar spray head* (which receives the mortar from the pump and sprays it out in a rain of mortar of the required fineness, velocity and spray cone angle)
The unit is sub-divided into:

- Spray head body
- Air jet
- Air jet valve
- Clamp for the air jet valve
- Renewable nozzle.

3.2.1 Chopper gun

As mentioned previously, the function of the chopper gun is to draw off the roving and cut it into uniform lengths which are then sprayed into the mortar spray by the air mover. The unit is driven by the chopper drive motor, which is usually an air motor.

Figure 3.13 shows the components of a typical chopper gun. The method of operation can be seen by referring to Figure 3.12. The polyurethane rubber drive

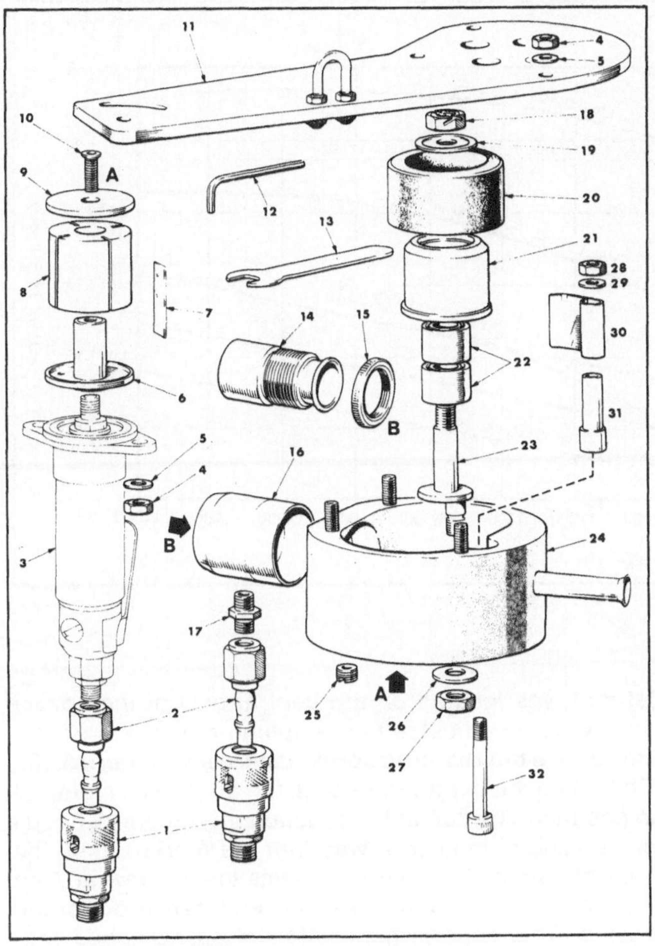

Item	Description	Qty
1	QR adaptor head	2
2	QR adaptor tail	2
3	Air motor c/w locking ring	1
4	Nut	5
5	Washer	5
6	Cutter block spindle	1
7	Cutting blades	4
8	Nylatron cutter block	1
9	Top cheek plate	1
10	Cutter block securing screw	1
11	Combination head plate	1
12	Allen key	1
13	Spanner for air motor shaft	1
14	Air mover inner	1
15	Air mover lock ring	1
16	Air mover body	1
17	Air mover air inlet fitting	1
18	Backing roller securing nut	1
19	Backing roller securing nut washer	1
20	Backing roller	1
21	Backing roller inner Nylatron	1
22	Backing roller bearings	2
23	Backing roller eccentric	1
24	Chopper body	1
25	Air mover securing grub screw	1
26	Backing roller adjusting nut washer	1
27	Backing roller adjusting nut	1
28	Tension spring retaining nut	1
29	Tension spring retaining nut washer	1
30	Tension spring	1
31	Tension spring tube	1
32	Tension spring bolt	1

Figure 3.13 Components of a typical chopper gun.

wheel (20) is set to run in contact with the cutting cylinder (8), which is driven by the air motor (3). The cutting cylinder has radial slots on its circumference which enable cutting blades to be housed (7). Once driven, the incoming rovings are chopped into lengths equal to the distance between the cutting blades measured around the circumference of the cutting cylinder. The blades are, therefore, spaced to produce the required fibre length. As the polyurethane roller (or backing roller) is driven by friction, the contact with the cutting cylinder enables it to support the rovings and allows them to be cut by a shearing action of the blade pressing radially or normally into the roving and backing roller surface. The setting of the contact pressure between the polyurethane roller and the cutting cylinder is critical. Too low a pressure prevents the roving from being cut and too high a pressure produces excessive drag on the air motor.

The air mover (14–16) is positioned at the mouth of the chopper gun and, by air pressure and air flow, draws the cut fibre out of the chopper gun and propels it into the mortar spray. Entry of the fibre into the mortar spray can sometimes be assisted by fixing a fan-tail nozzle to the outlet of the air mover. This fans out the fibre and therefore increases the amount that can be taken into the mortar and also the rate at which it can be coated with the mortar. Figure 3.14 shows a fan-tail nozzle fitted to the chopper gun for use on a proprietary traversing spray-up head unit.

The use of an air motor to drive the chopper gun is governed by the allowable weight limit. For hand-spray work it seems that the power output from an air motor is ideally suited, particularly when one considers that the unit has to be man-handled for an 8-hour shift. Electric motors would give more consistent outputs of fibre and are easier to set and correct but their excessive weight and safety considerations have prevented their use. However, for automated production and traversing head units, they would offer great benefits.

The most common air motor used for driving grc chopper guns is the Desoutter TM Series. This is a 0.22 hp motor offered with three speed settings 3000 rpm, 4000 rpm or 5500 rpm. Figure 3.15 shows the performance for this motor. It can be seen that maximum torque and brake horse power occur at rotator speeds of 17 000 rpm and less for the indicated operating pressures. This means that when a set gearing from rotor to final output shaft is employed (usually 7.7:1), the

Figure 3.14 A transverse head, spray-up head unit, illustrating a fan-tail chopper gun nozzle.

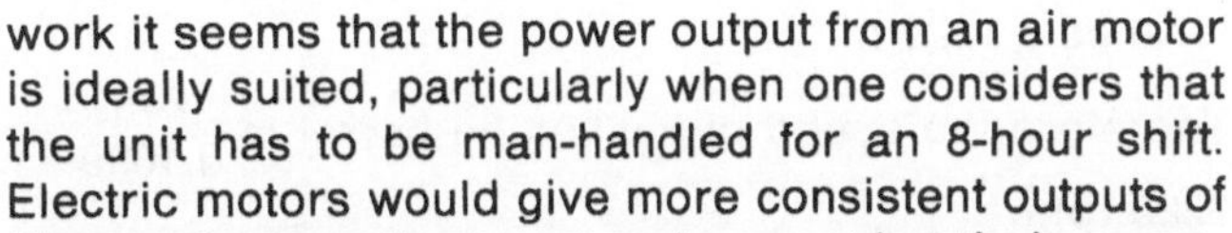

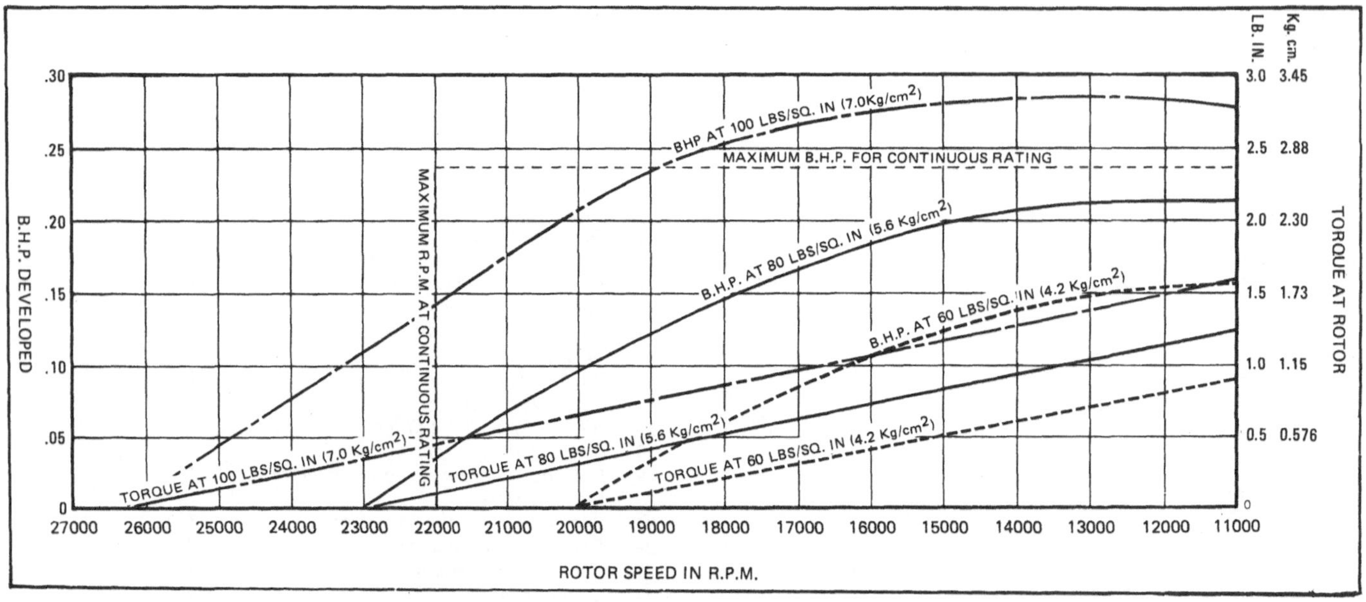

Figure 3.15 Graph showing the power developed by the TM Desoutter air motor at the speed and air pressures quoted.

speed of the chopper gun would therefore be fixed at a maximum. This maximum should not be exceeded if optimum power is to be available to cope with roving friction and varying loadings.

Example:

If a typical composite output rate of 11 kg/min. is taken this would employ 600 g/m of chopped fibre at 5.45% fibre content.

At a roving weight of 2.45 g/m this represents 245 m/min. $\frac{600}{2.45}$.

The diameter of the cutter barrel is 35 mm, the circumference 0.1099 m.

Therefore the speed of the cutting barrel is $\frac{245}{0.1099}$ = 2230 rpm.

Rotor rpm = 2230 × 7.7 = 17.171 rpm.

The chopper gun is thus being driven at a motor speed within the ideal range for optimum performance.

If, however, the composite output is increased to say 15.85 kg/min. and the same conditions exist as before, a cutter barrel speed of 3200 rpm would be required and consequently a rotor speed for the air motor of 24 640 rpm which is way off the optimum running speed.

This factor can account for many apparent fluctuations in the output of air driven chopper guns and requires either:

1. Modifying the gearing of the output shaft

or

2. Changing to a more powerful motor.

3.2.2 Mortar spray head

The function of the mortar spray head is to receive the pumped mortar and transform it into a fine rain of the required cone angle β (Figure 3.12). This is achieved by propelling the mortar through a nozzle at a velocity increased by a jet of compressed air. Both the increased velocity at the nozzle and the pressure drop across the nozzle as the mortar escapes are regulated to produce the required cone angle and velocity.

Early production models of spray heads used an air jet that was located on the central axis of the nozzle and could be adjusted to the optimum distance from it to produce the required spray pattern (see Figure 3.16). On finding the optimum setting, the air jet was clamped in position. The current production mortar spray heads are arranged in such a way that the mortar takes the central path and the air jet directs the air flow around it in a concentric spiral manner and hence out of the nozzle. This second generation of spray heads produces a far superior spray rain of mortar and has the

Figure 3.16 Cross-section of an early production mortar spray head.

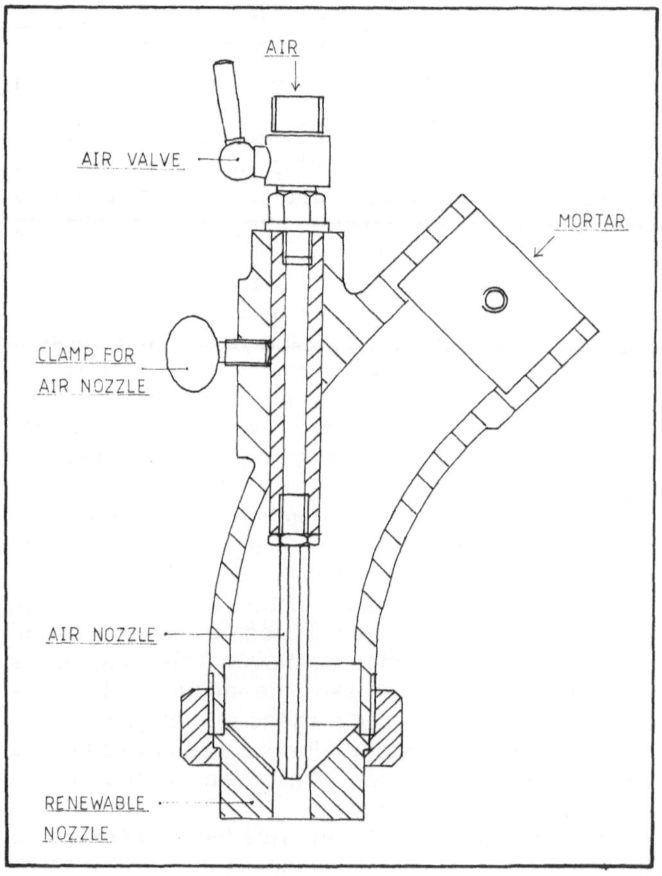

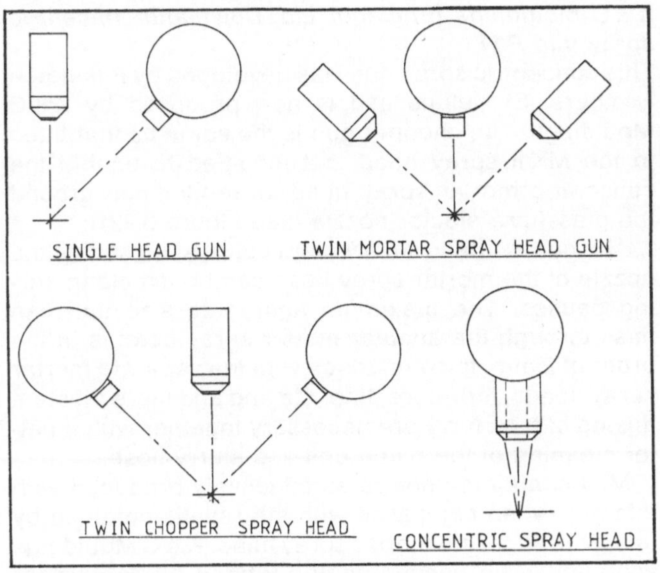

Figure 3.17 Basic spray head types available.

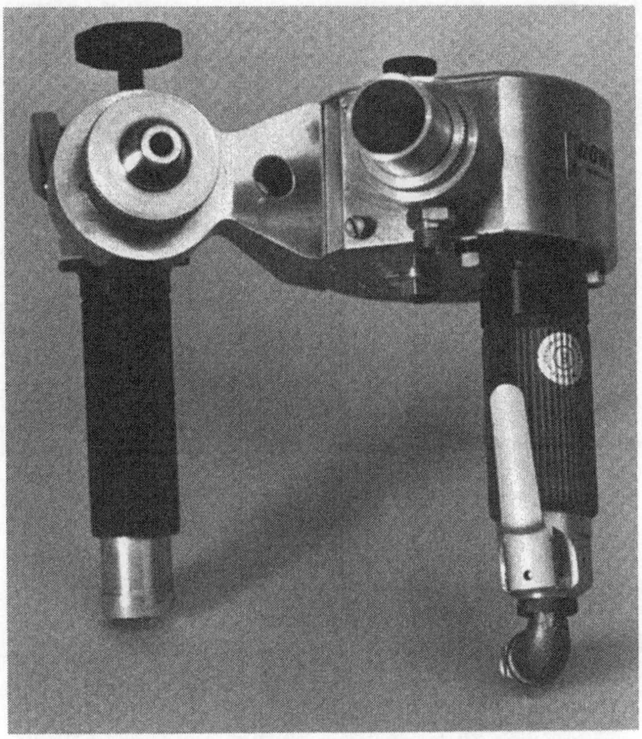

Figure 3.18 Downland separate single spray head.

advantage of fixed jets with no setting problems or errors.

3.2.3 Twin mortar head spray head

Another early development was the twin mortar head spray head. This employed two standard spray heads located either side of a chopper gun. It has again proved successful but suffers from being a heavy spray head for manual work. As described earlier, if chopping outputs are increased to match the higher mortar output obtained from this type of spray head, the chopper gun motor must be checked to ensure it is not operating outside its optimal power range.

Consideration has been given to the system and some work has been carried out using twin chopper guns located each side of a mortar spray but no data on performance are available. Figure 3.17 shows the basic types of spray head configuration.

3.2.4 Concentric spray head

Recently a new design of spray head has been developed known as the concentric spray head. This combines the mortar spray head with the chopper gun by locating the mortar spray head around the nozzle of the chopper gun. The chopped fibre is then sprayed out in the centre of a concentric ring of mortar. This spray head produces a good spray pattern that 'contains' the fibre and considerably reduces wastage. Its disadvantage is a tendency to block easily if particles of sand or cement greater than 1 mm are present in the mortar. However, it is proving to be a viable step forward in spray head design.

3.2.5 Spray heads and their manufacturers

Detailed below are some of the types of spray heads that are currently available or have been produced for grc production.

K & C Mouldings (England) Ltd, separate single spray head

This unit consists of the Downland chopper gun and the MK II mortar/slurry gun. The two heads are located to produce a fixed inclusion angle α by means of a casting bolted to the chopper gun and clamped around the nozzle of the mortar gun. The mortar gun can be twisted to produce a comfortable 'hold' and then clamped in position. Figure 3.18 illustrates the spray head with the clamp screw securing its position and also the lever arm for the air mortar control to operate the chopper gun.

The air supply to the mortar head has a shut-off valve which is operated by one's thumb. This valve not only stops the air supply to the mortar head but also, by increasing the back pressure when closed, is arranged to shut-off the mortar pump and therefore the supply of mortar to the spray head.

The spray head can be suspended by attachment to a hole provided in the depositor bracket.

The cutting cylinder in the chopper gun can be set to produce chopped fibres of either 18 mm length or 37 mm length. The former length is achieved with six blades in the cylinder and the latter length with three blades. The blades are equally spaced around the circumference in each case.

Settings for the K & C Mouldings separate single spray head are given in Table 3.5.

Table 3.5 Settings for the K & C Mouldings separate single spray head.

Air mover	20–60 psi (dependent on the glass output required) 5–6 cfm at 40 psi
Chopper gun	50 psi
Mortar spray	10–20 psi

NOTE:
The air feed to the chopper gun air motor must be taken through an airline lubricator.

Power-Sprays Ltd, separate single spray head

This unit consists of a standard chopper gun unit and

Figure 3.19 Power-Sprays Ltd separate single spray head.

Figure 3.20 Putzmeister twin mortar spray head.

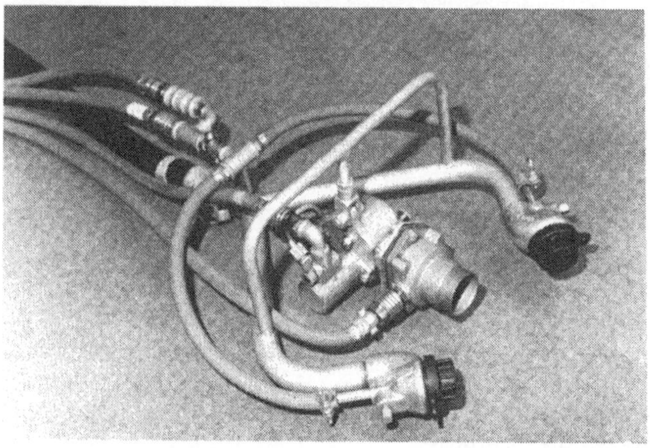

the CEM-TEX mortar spray gun. Its distinguishing feature is a rotary spool valve that produces an instant shut-off of the ejecting mortar which closes down the pump. Figure 3.19 shows the spray head unit with the plate mounting for the chopper gun and the mortar spray head. The adjustment for the inclusion angle α can be seen as well as the operating lever for the shut-off spool valve (above the operator's left thumb).

Putzmeister twin mortar spray head

Although somewhat overshadowed by the development of the concentric spray head, the twin mortar spray head has been used by manufacturers to speed up outputs and to ensure a good intermixing of mortar and fibre lay-up. The Putzmeister version uses mortar nozzles taken from plastering machines. They are made of rubber which allows for rapid cleaning and reduces the likelihood of blockage due to oversize particles in the mortar. The whole unit is rather heavy and bulky but produces the desired effect (Figure 3.20).

Drostholm multimix concentric spray head

Concentric spray heads are possibly the best method of spraying grc. The Drostholm company led the way in developing this technique but have now, unfortunately, left the product scene. The principle used, however, has been taken up by other manufacturers.

The Drostholm gun was a heavy unit and rather cumbersome in operation. The rights of manufacture are now with Power-Sprays Ltd (Figure 3.21).

K & C Mouldings (England) Ltd, Downland concentric spray gun P 11

This concentric spray gun was developed by Pilkington Brothers, St Helens and is now produced by K & C Mouldings. The chopper gun is the same as that fitted to the MK II spray head, but modified to enable the concentric mortar spray head to be clamped around the glassfibre ejector nozzle (see Figure 3.22).

The gun is fitted with a water flush system so that the nozzle of the mortar spray head can be left clean during pauses. The maximum aggregate size that can pass through the annular mortar spray head is in the order of 2 mm. In comparison with the separate mortar spray heads, therefore, finer sieving and more efficient mixing of the slurry are necessary together with a better cleaning of the pump unit and slurry hose.

Mist coats may not be as effectively produced with this gun when compared with the results obtained by using the separate mortar spray head. K & C Mouldings will supply the concentric spray head on its own for manufacturers to fit to their existing MK II Downland choppers.

The specification for the Downland concentric spray gun P 11 is given in Table 3.6.

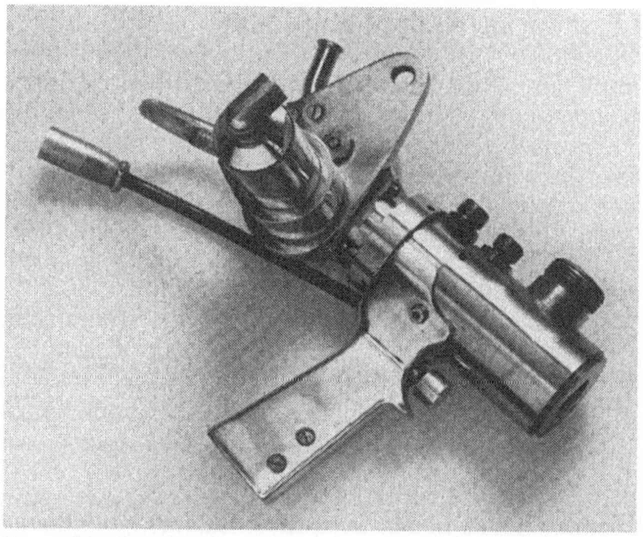

Figure 3.21 Drostholm multimix concentric spray head.

Figure 3.22 Downland grc P 11 concentric spray head.

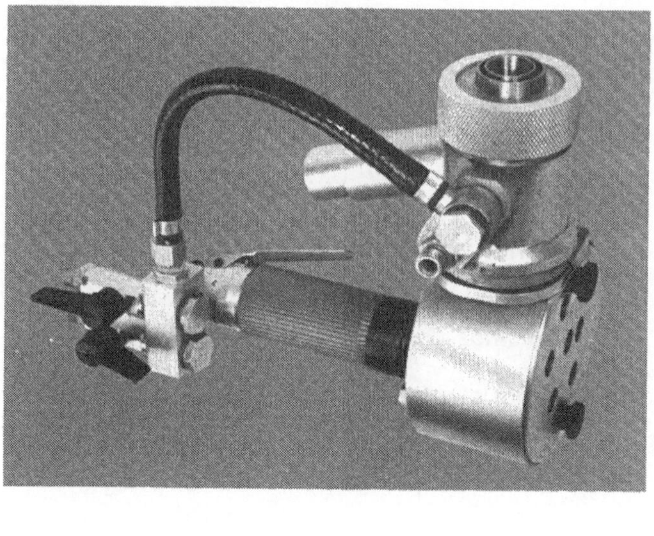

Table 3.6 Specification for the Downland concentric spray gun P 11.

Air consumption		50 cfm
Maximum glass output:		
	3000 rpm	750 g/min.
	4000 rpm	1000 g/min.
	2000 rpm	500 g/min.
Glass strand lengths (this is with 6 slot cutting cylinders 12 slot are also available)		18 or 37 mm
Maximum slurry output		20 kg/min.
Total weight without hose		2.7 kg

The concentric spray gun can be regarded as three parts in one. The inner section consists of a main tube and fibre inlet tube shown as (5) and (3) Figure 3.23. It is basically a venturi supplied with compressed air to create the main air stream of the gun. The stream sucks the chopped fibre through the gun and blows it into the surrounding slurry. Fibre tubes fit into the spray body (7) forming the second part. A chamber in between them directs the slurry into an annulus around the fibre tubes. The third part is used to break up the slurry stream and direct it into the air stream. It consists of a chamber formed by the spray body, two slurry caps and the retaining ring (7), (10), (11) and (13). The chamber is fed with compressed air which escapes through the serrations between the two slurry caps and into the annulus. It is also supplied with water to enable the nozzle of the gun to be kept clean during pauses in spraying.

The gun should be assembled as shown in Figure 3.23, the components having been slightly greased with a silicon based lubricant. The pressure settings required are:

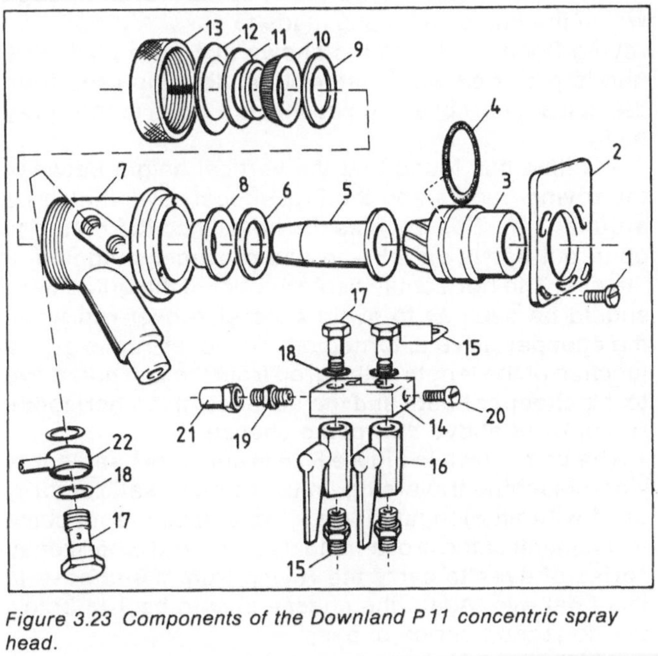

Figure 3.23 Components of the Downland P11 concentric spray head.

(a) Fibre air pressure 30 psi (2.1 bars)

(b) Slurry air pressure 45 psi (3.2 bars)

(c) Air pressure for the chopper motor should be set to give a glass output of 630 g/min. This will have to be re-set for different rovings, or if the backing roller is moved. Rovings must not be chopped without the air mover in operation

(d) The pump output should be set to 12 kg/min. output.

Again, the atomizing air must accompany the mortar output. The nozzle must be flushed with water after the pump is switched off.

Putzmeister concentric spray head

The Putzmeister concentric spray head is illustrated in Figure 3.24. A cross-section diagram (Figure 3.25) gives the various parts of the unit which can then be seen in operation. This spray head has a rubber wear-resistant nozzle and central integrated fibreglass chopper. The nozzle pipe is adjustable and allows for optimum setting of the mortar spray.

The unit permits aggregates up to 3 mm to be processed without blockage. When necessary the spray jet or mortar cone angle β can be narrowed for covering small surfaces or widened for spraying over large areas. Output is 30 litres/min.

Power-Sprays Ltd concentric spray head

This is the most recent concentric spray head to come onto the market and as yet little information is available on its performance. It is marketed as suitable for normal grc as well as polymer modified grc composites.

K & C Mouldings (England) Ltd, Downland grc auto head

This unit (Figure 3.14) incorporates the Downland glass chopper and MK II mortar spray head mounted on a crossbar for attachment to the traversing unit on automated plants.

The mortar spray head uses standard parts but the chopper head is fitted with a more powerful air motor to enable higher glass outputs to be maintained. A larger air motor and fishtail nozzle gives a more controlled placing of the fibre output.

The specification for the Downland grc auto head is given in Table 3.7.

Table 3.7 Specification for the Downland grc auto head.

Total air consumption	35 cfm
Maximum output:glass	1.3 kg/min.
Maximum output:mortar	20 kg
Weight	6 kg

ARC continuous roving spray head

This spray head has been developed for the production of grc concrete pipes but, although only used commercially in manufacturing these units, the principle could be well adapted to produce other more general components.

The spray head works on the basis of drawing off roving and projecting them in a sine wave pattern onto a revolving mould face. The roving is not cut but left as

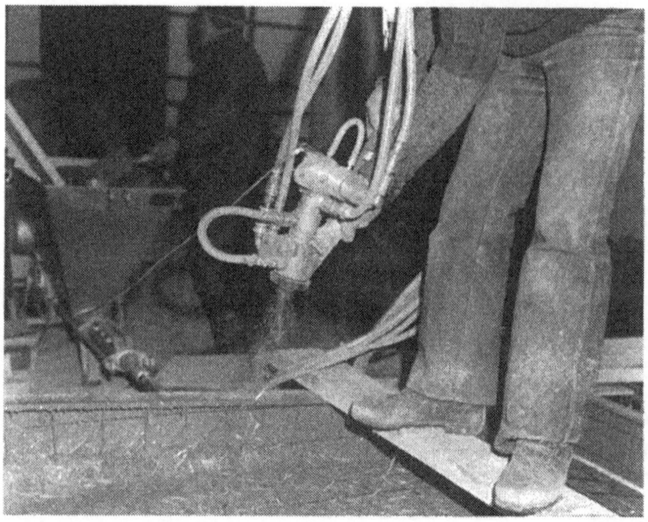

Figure 3.24 Putzmeister concentric spray gun.

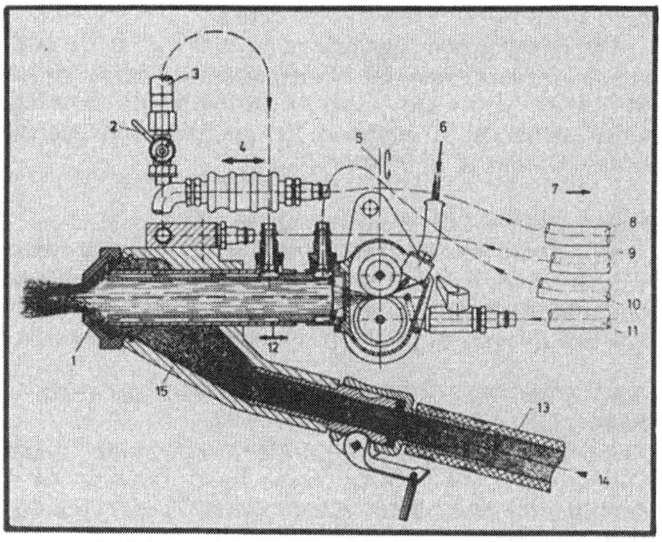

Figure 3.25 Components of the Putzmeister concentric spray gun.

1 Rubber nozzle
2 Cock to regulate air for spraying
3 Air for spraying
4 On/off
5 Fibreglass cutter rotates on own axis
6 Inlet for fibreglass rovings
7 To air distributor
8 Control air
9 Air connection
10 Air nozzle
11 Air for fibreglass cutter
12 Adjustable
13 Connection to mortar pump
14 Mortar
15 Nozzle neck can be adjusted for adaptation to various aggregate sizes and to reduce or increase the spraying angle.

a continuous strand. The spray head traverses at right angles to the direction of mould travel to produce a winding type of lay-up. Mortar is sprayed simultaneously.

The sine wave lay-up pattern produces considerable benefits in efficiency as can be seen in the section devoted to this process further on in this Chapter.

3.2.6 Further notes on spray heads

It should be noted that the grc chopper spray head has come from the grp industry. The basic grc chopper gun is different only in that the backing roller is produced from a polyurethane rubber with a shore hardness of 95. This means that other chopper guns used for grp could be adapted for grc production.

It is clear that more down time is directly the result of problems with the chopper gun, than any other source. Consistent output is also of concern since a small change in the rate of output from the chopper gun has a significant and problematic effect on the fibre content. Again, it should be remembered that the air motor driven chopper gun is a development from the grp industry; where glass content of 50% is required and a $\pm 2\%$ variation has little significance. For grc production, however, such variations are critical.

Hopefully a new generation of chopper guns will be developed that utilize a more consistent power source than that offered by air motors, will not slip when taking up a roving and employs a better cutting action than at present.

Before leaving the subject of spray heads, it should be mentioned that premix grc has been shown to be sprayable. This technique cannot produce composites of 2D lay-up pattern and, therefore, will not have properties in the same class as pure spray-up grc. Nevertheless, by spraying pre-mix, it is possible to use this technique as a placing method. Various adaptions have been employed — namely screw feed pumps and concrete placer type machines — as the motive force together with conventional plastering nozzles and straight through nozzles with atomizing air assistance. However, the fibre content has not been great and the fibre length around 12 mm. Often admixtures have been required to assist the flow through the pump and to the spray head. This system may well be useful as a site technique where spraying premix could be ideal for block bond application or fire wall and ground stabilization uses.

3.3 ROVING FEED SYSTEMS

In order that the wound package forming the roving cheese can be efficiently fed to the chopper head, a suitable 'path' has to be established. This has evolved resembling the fishing rod method of supporting the fishing line—namely a series of spaced eyes through which the line or roving is made to pass. A production roving boom and stand is shown in Figure 3.26. This unit is provided with castors to enable the drop from the last support eye to be positioned close to the spray head.

Practice has found that the vertical height between the roving bundle and the first support eye should be around 1 m. This enables the heavily coiled roving to open out easily on take-up and prevents snagging at the eye. The horizontal displacement of the guide eyes should be such as to avoid excessive over-run when the chopper drive is turned off. This over-run is also a function of the length of the drop from the last guide eye to the chopper head, and the distance of the horizontal run of eyes above the roving cheese.

The unit shown in Figure 3.26 is suited to hand-spray work. Machine traversing units are more satisfactorily used with an elongated eye lying across the machine in the plane of the cross-head traverse and a horizontal series of eyes to carry the roving from the cheese to this final eye. Again the cheese should be 1 m below the horizontal series of eyes.

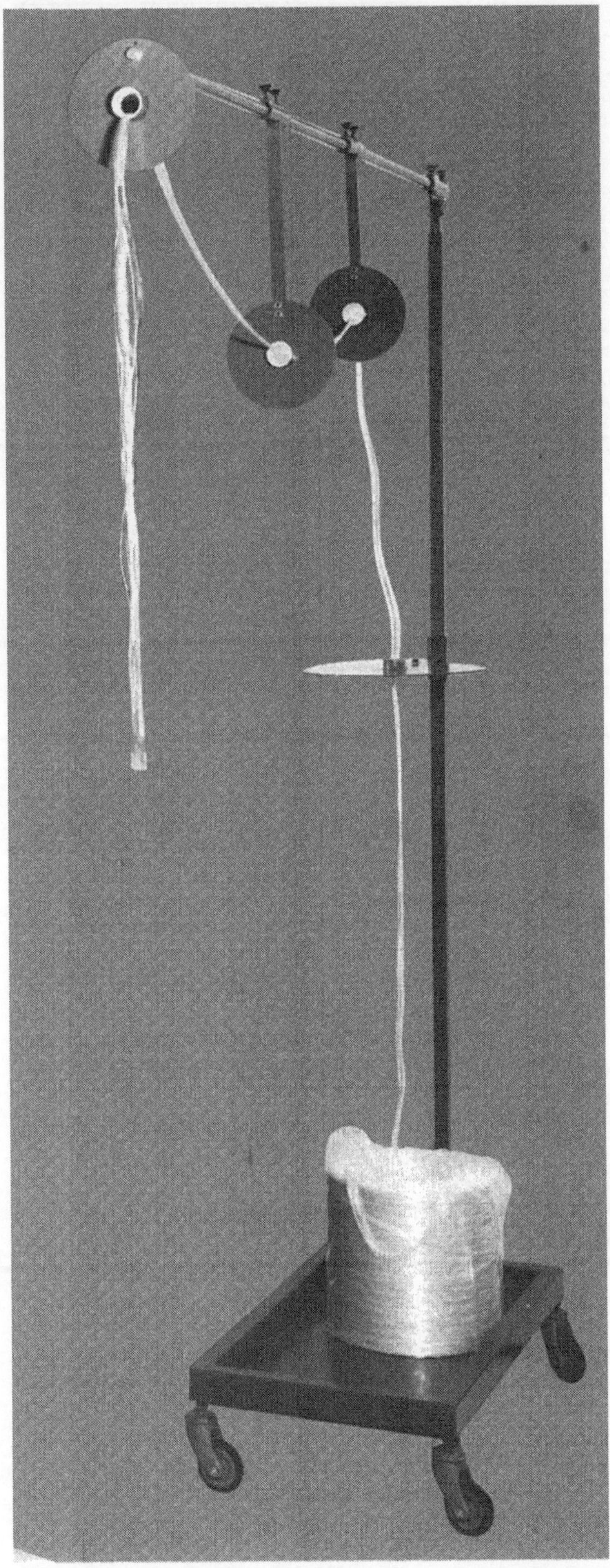

Figure 3.26 A production roving boom.

The eyes are usually made to contain a ceramic ring which has a high abrasion resistance and a low frictional resistance on the roving. This prevents chaffing of the roving. An internal diameter of circa 25 mm is adequate. The elongated eye, used as the final eye set across the traverse head unit, would extend almost the entire width of the traverse path and have a vertical aperture of at least 200 mm. A stainless steel rod is ideal for this component.

3.4 WATER GAUGING EQUIPMENT

So often it is found that production set-ups are carefully organized for the actual manufacture or spraying and handling of the composite, but the simple process of gauging the water content of the mix is left as a bucket or hose pipe and eye assessment. The expert batcherman may well produce a good, consistent end product, but scarcity of experienced labour can easily result in the ruin of a potentially good grc by excessive additions of gauging water.

It is, therefore, crucial that the batching process should include a device for dispersing accurately into the mix the quantity of water required. Such devices can be obtained commercially or manufactured in house. Flowmeter dispensing is not preferred since variations easily occur.

A simple and effective unit can be produced from using a lavatory cistern-type ball valve attached to a vertical slide and housed in a calibrated tank. The quality control operative would then be responsible for setting the batch quantity of the water and noting this in the daily quality control diary.

3.5 SUGGESTED MOBILE RIG FOR GRC PRODUCTION

In the author's opinion production of grc has now reached a stage where the basic 'know how' for producing 'good' grc exists. What is lacking is cost effective methods. The production unit needs flexibility. It may be required, for example, to provide white grc for some part of the shift and then standard oPc grc for the remainder. Equally, grc could be just one form of precast concrete that a precaster is using in his day-to-day production. A suggested solution is the introduction of batch production as orders are received with a reversion to other methods of precasting in between. Alternatively, the situation may arise where the mould would be best left stationary and the production capacity moved along or through the factory.

The facility in Figure 3.27 evolved to satisfy the above aims. This unit consists of a rigid frame on castors. To this frame is positioned the pump unit and mixer. The latter is equipped to discharge straight into the hopper of the mortar pump. An articulated boom is fitted to the frame to carry the weight of the spray head and act as a carrier for the mortar hose from the pump to the spray head. The roving is also fed along the boom through eyes. Sand and cement are supplied in pre-gauged quantities to suit the batch weights per mix. Water is gauged by a cistern, as described earlier, and the whole rig is wired and plumbed so that single connections to an air, water and electricity supply are all that is required.

The roving is housed in a unit that also contains a facility mentioned in *Chapter 10* for continuously monitoring the fibre inclusion percentage. This is noted on a pen recorder either housed on the unit or at a separate location. The on-going output is displayed via an analogue print-out as a percentage fibre inclusion and can be set or adjusted by the batcher, with controls to vary the mortar or chopped fibre outputs.

This unit would offer a complete spray facility ready for connection to the services and could therefore be ideally suited for export as a total spray production unit.

3.6 VACUUM EQUIPMENT

Vacuum extraction of surplus water from 'green' grc

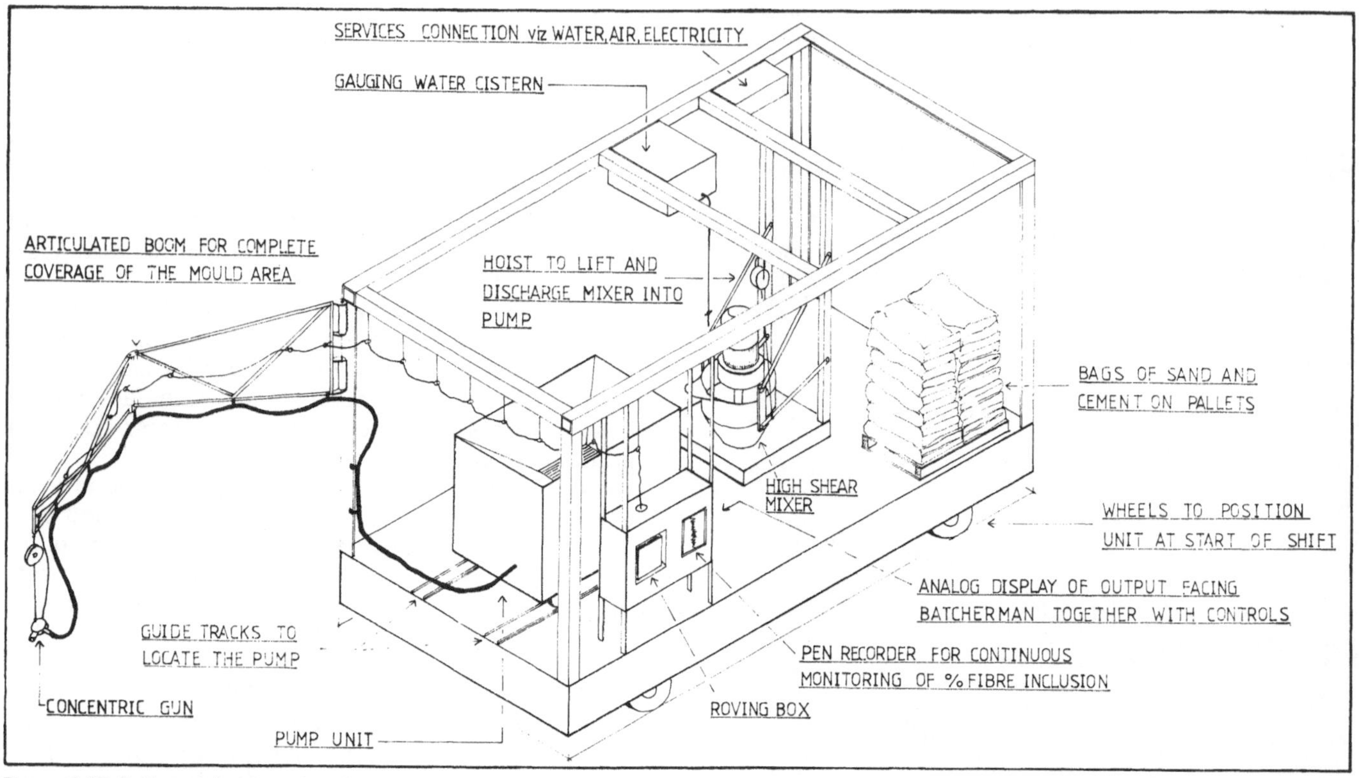

Figure 3.27 Self contained grc hand-spray facility.

components has certain advantages in the improvements it produces. Shrinkage of the composite is reduced and the strength properties are improved due to the lower water/cement ratio and increased density. Production control is eased because the gauging water addition is not so critical—the surplus will be extracted in any case.

Early development work praised the benefits of dewatered grc composites, but production was limited to mainly flat sheet or post spray forming. Upward vacuum extraction has shown, however, that patterns and a limited range of profiles can be produced.

3.6.1 Downward vacuum extraction equipment

This can take the form of either a stage or continuous process, but both operate in the same manner.

A shallow box of suitable dimensions is produced with air-tight sides and base but a stiffened top that contains a grid of holes at approximately 10 mm centres and of circa 4 mm diameter. The stage process then requires the top to be covered with a filter paper or fabric. Grc is sprayed onto this surface until the required thickness is reached and trowelling gives a uniform finish. The box is connected to a vacuum pump and is separated by a filter system. Vacuuming commences and the time and degree of vacuum is fixed by experiment. Surplus water is tapped out of the system and the vacuum removed. Finally, the board is taken off by either suction lifts or inversion of the box.

The continuous process differs only in that the conveyor belt is arranged to carry wet composite over the vacuum box and the vacuum stage is finalized before the composite has completed the traverse across the vacuum box.

3.6.2 Upward vacuum extraction equipment

This process is different because grc is sprayed onto the mould or profiled plate (only shallow profiles are possible), and then a filter cloth or paper is placed onto the top of the wet composite. Following this the vacuum box is positioned over the top of the composite with the perforated plate against the filter material. Air is extracted and, provided an air-tight seal is obtained around the perimeter of the mould and the vacuum box, the mould is sucked against the box. This sandwiches the grc which, therefore, takes on the profile of the mould. Hence water is extracted by pressing and vacuum suction.

Equipment of this type can produce detailed finishes — low relief or smooth matt surfaces on the mould face and good general finishes on the back.

With both the downward and upward techniques, virtually flat products are possible. The vacuum extraction method has not, however, been particularly popular because it introduces a further stage into the production process and requires a large capital outlay. Also, the detailed mouldings which form the majority of the production of grc, to-date, would require expensive moulds if they were to incorporate dewatering vacuum extraction. Small compact versions of the upward technique, about 300 mm^2, have been used as a hand held tool for removing excess water from wet composites in profiled mouldings but its use remains limited.

3.7 AUTOMATED SPRAY-UP EQUIPMENT FOR SHEET PRODUCTION

The spray-up process can be mechanized to produce flat sheets or shallow profiled products. The functions normally carried out by the spray operative are taken over by a traverse unit. This unit holds the spray head assembly (of the type shown in Figure 3.14) and passes it across the mould. There are two systems using this technique:

- Continuous plant process
- Station plant process.

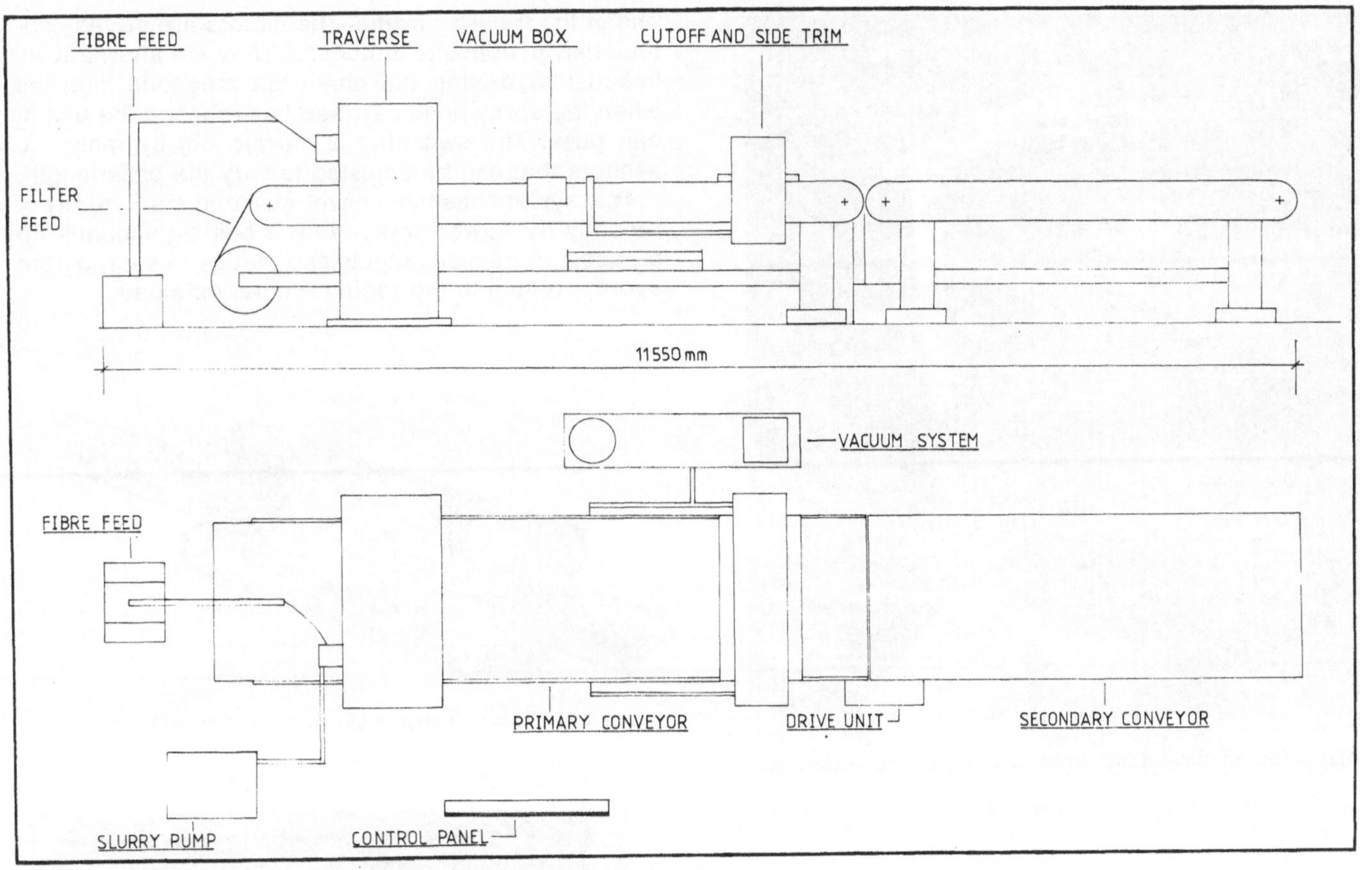

Figure 3.28 Procem continuous dewatered flat sheet production plant.

3.7.1 Continuous plant process

This particular process has led to the development of the Compo-Cem flat sheet machine and the Procem flat sheet plant (Figure 3.28). Both processes employ dewatering, and are basically very similar. The Procem plant is described here.

The conventional mortar pump and spray head arrangement is employed and the process initiated by the traverse unit (described later in this Chapter). Grc is sprayed onto a primary conveyor, which is an open weave synthetic material capable of permitting water to pass through during the vacuum process. The primary conveyor receives the spray of grc and a continuous lay-up of uniform thickness is achieved by adjustment of its forward motion together with the cross traverse motion of the spray head. Setting these two rates of traverse is critical to ensure a uniform board thickness. The primary conveyor passes over the vacuum box where excess water is sucked downwards and out of the wet composite. This is performed by a liquid ring vacuum pump. After vacuum treatment, the wet board is cut and side trimmed. The cross cut is achieved by a flying shear type cross cutter which travels automatically with the board so as to produce a square cut. Two adjustable side trims cut the board to the required width. An automatic length selector initiates the moving carriage cross cutter to cut the board to the finished length. A secondary conveyor is driven by the primary conveyor and serves to remove the finished board ready to be stacked for curing.

The primary conveyor belt is fed to the process by a unit located at the front of the plant. This production unit is designed to produce flat sheet grc at variable widths up to 1.5 m wide in thicknesses from 3 mm to 12 mm.

The Compo-Cem unit is operated by Compo-Cem Ltd, a joint company of Tunnel Holdings Ltd and Pilkingtons.

The Procem plant is the process manufactured and marketed by Procem Company Ltd, London Road, Hazel Grove, Stockport, Cheshire

3.7.2 Station flat sheet plant

To achieve the production of a flat sheet of grc this process involves four stages with a table transferring the composite from one stage to the next. The sheet is sprayed-up by continuous sweeps of the spray head mounted on a traverse unit which progresses along the mould table. As the table moves from station 1 to 2 the board is formed. Dewatering takes place at station 3. Finally, at station 4, the sheet is removed from the table and transported by vacuum lifts or transfer machines developed for the purpose. The boards are then stacked on pallets which are built up 12 high and transferred as a unit for overnight curing. The production rate is one 3 m × 1.2 m × 6 mm grc sheet every 5 minutes. Figure 3.29 illustrates the basic machine undergoing proving trials.

This machine has been developed by Stelmo International, Westwell Leacon, Ashford, Kent.

3.8 AUTOMATED TRAVERSE UNITS

The two systems outlined above rely on automatic traverse units to convey the spray head across the mould at a constant set rate. So far three methods have been produced to drive the traverse arrangement:

- Mechanical traverse system
- Enclosed air plug or linear air motor system
- Constant speed electrical drive system.

Figure 3.29 Stelmo station flat sheet plant undergoing proving trials.

3.8.1 Mechanical traverse system

This system employs a variable speed electric drive motor to drive a continuous cable. The necessary oscillation is achieved by reversing switches situated at each end of the traverse and reversing the cross head by changing the polarity of the supply to the drive motor.

3.8.2 Enclosed air plug or linear air motor system

This system has proved admirable and works well in the dusty environment of grc production. A complete set-up as marketed by Power-Sprays Ltd, uses the Mardrive drive unit, a cross section of which can be seen in Figure 3.30.

The air plug is enclosed in the seamless aluminium cylinder tube which it crosses under the force of air pressure. Velocity control is achieved by an air bleed valve located at the far end of the table towards which the plug is moving. The drive to the carriage, which is supported under the aluminium cross tube on an RSJ, is via permanent magnets located within the air plug. They create a magnetic field that draws an externally moulded carriage slung around the aluminium air tube which is then connected to the spray head carriage. Figure 3.31 illustrates the unit in operation. Below this can be seen the RSJ carrying the spray head and taking its weight. Consequently, the air plug has only to overcome the friction of the external carriage and the spray head carriage running on the RSJ. Reversing the drive at each end of the traverse is achieved by air switches changing over the air supply and bleed valve arrangement to reverse the motion.

3.8.3 Constant speed electrical drive system

The principle of this system is to drive a continuous chain linked to a spray head carriage (as shown in Figure 3.31) via two electric motors situated at each end of the traverse frame. The motors are driven continuously in opposite directions. They are alternatively linked into driving the chain via magnetic clutches when the spray head carriage has reached the end of one pass. The switching is carried out by magnetic sensors that can be adjusted to vary the pass length.

This system has been developed and marketed very recently by Power-Sprays Ltd. It has significantly increased production outputs and reduced wastage from overspray due to the rapid reversal obtained.

Figure 3.30 Section through a Mardrive traverse unit.

Figure 3.31 Mardrive traverse unit in operation.

3.9 FLAT SHEET MASS PRODUCTION PLANT

The Hatschek asbestos cement sheet process has received attention as a production method suitable for large outputs of grc sheet. However, if it is to produce grc, adaptions are required costing about 20% of the price of a new plant. These include modifications to the feed stock preparation and also the sheet handling and curing areas, as well as the plant itself. Such adaptions would not affect the output of the machine and equal outputs are obtained on both products (typically 5 tph).

A typical Hatschek process is shown in Figures 3.32 and 3.33 and can be summarized as follows:

1. Cement, asbestos and pulp are put into the mixers and blended with water
2. Slurry from (1) is stored in the chest tank
3. Water from the header tank is added to the slurry to adjust concentration. The slurry is then fed to the vats
4. Slurry is filtered in the vat by a wire cylinder, then the removed solid is transferred to an endless felt
5. Wet film is dewatered and transferred to the making roll to be rolled until the required thickness is reached

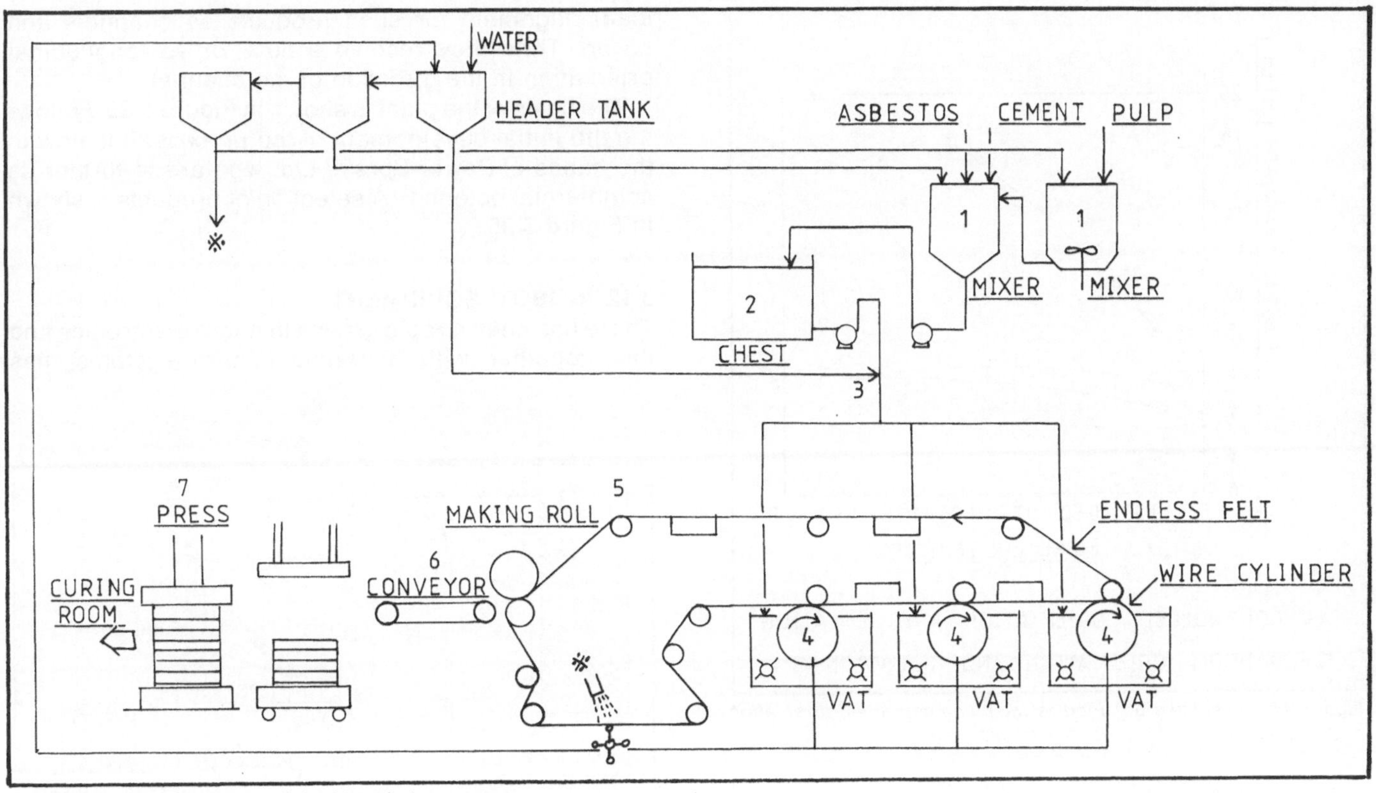

Figure 3.32 Flow diagram of the Hatschek process.

Figure 3.33 Hatschek plant developed by TAC Construction Materials Ltd.

6. The roll formed at (5) is cut and transferred to the conveyor. It is now called the green sheet
7. After pressing, if required, the green sheet is taken to the curing room.

This process relies on the adhesion of asbestos to the cement slurry during dewatering and transfer from stage to stage. When glassfibre is processed, adhesion is much reduced and therefore larger quantities of recycled water and cement solids have to be handled. Interlamina strength of the wet film is poor when using glassfibre and can lead to delamination and lack of transfer during production. Both the pulping and mixing stages tend to defilamentize the glassfibre strands causing intertwining and irregular distribution of the fibres, and therefore variations in reinforcing effect within the board.

With these difficulties to overcome, the TAC Construction Materials Ltd plant and the Asahi Glass Co. Ltd in conjunction with Sanno Ltd have patented solutions. The TAC Board is produced without asbestos in the composite, but uses various fillers to assist production. The Asahi Glass Co. Ltd, with Suncrete Board, uses additions of asbestos limited to below 5%.

The TAC Board claims a minimum bending strength of 15 N/mm² at a minimum density of 1300 kg/m³ while the Suncrete Board claims, with asbestos, a minimum bending stress at failure of 24 N/mm² at a density of 1700–1800 kg/m³. Development is continuing.

3.10 ARC SLIMLINE GRC PIPE MANUFACTURING PLANT

The ARC Group has developed a production plant that produces a concrete pipe suitable for sewer and surface water applications. This process replaces the wire reinforcing cage that is used in conventional pipes, with an inner and outer skin of less than 1% of direction-controlled glassfibre roving. Pipes in production are in the range of 600 mm to 1.2 m diameter.

After injection of the helically wound glassfibre roving and mortar spray, the body concrete is introduced, vibrated and centrifugally dewatered. The internal grc layer is then formed, followed by the introduction of fine concrete to make the integral lining. The pipes are finally steam cured and allowed to mature for at least 4 weeks before delivery.

Probably the most interesting aspect of this process is the effect achieved by using the special fibre feeder to place the continuous roving of glassfibre and its correspondingly dramatic efficient usage of the roving as reinforcement. Figure 3.34 shows the relationship between fibre content and flexural strength for this placement method in comparison with random orientation and location. The addition of 1% of controlled orientation and location of continuous roving trebles

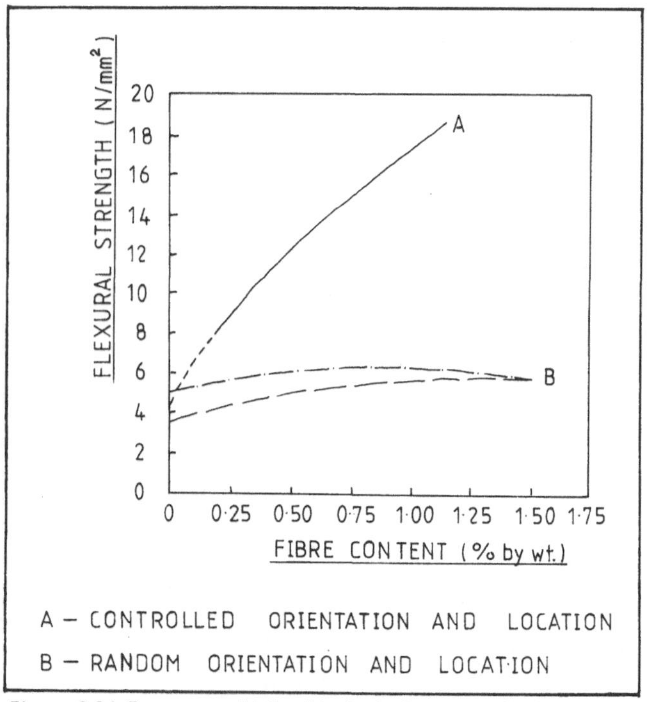

Figure 3.34 Response obtained by including controlled orientation and location of glassfibre rovings to concrete.

the flexural strength to that of random placement methods. These results given in Figure 3.34 are average results, but the coefficient of variation was significantly less with the more accurate fibre control achieved by this process.

3.11 BANCEM EXTRUSION CONTINUOUS PROCESS

The combination of high shear mixed mortar and the controlled application of vibration within the premixed composite enables sufficient fluidity to be obtained to operate an extrusion process for wet grc. This was the concept of the Bancem process.

The process consists of conventional silos with automatic feed to a blending system which supplies a regular premix of solids into the mixing unit. The mixing unit is capable of producing in excess of 200 kg of sand/cement slurry per minute and can accept perlite, styropor and other fillers. Normally the water/cement ratio is below 0.3 and, due to the high shear action, is in a fluid state. At this point the fibre is introduced direct from a chopper unit blending with the matrix material in the mixer unit and falling into the machine hopper.

Internal vibration at selected frequency and amplitude is applied. This causes intimate integration of fibre and matrix material and fluidizing of the mix to flow into the receiving moulds travelling through the machine. The vibrator consists of a horizontal tube containing an eccentric mass and is positioned across the machine in the fibre mixing hopper. The moulds are supported in carriers positively driven through the machine at a controlled speed. A vibrating screed plate surfaces the material, and the moulds and carriers continue through the plant where the moulds are removed by vacuum lift and stacked for curing after a short period of setting and hardening. The mould carriers are then returned to the system for recirculation.

Vibration of the premixed composite enables fibres to align into a 2D orientation, thereby achieving the ideal alignment for such products as channels and so on. Then they receive a 60% or so longitudinal orientation in the direction of the channel.

The basis of the plant is shown in Figure 3.35. Following the initial development of the process, it is now in the hands of Power-Sprays Ltd, who are to further its commercial potential. A selection of products is shown in Figure 3.36.

3.12 'ROBOT' EQUIPMENT

There has been a rapid growth in micro electronics and this, together with hydraulic control systems, has

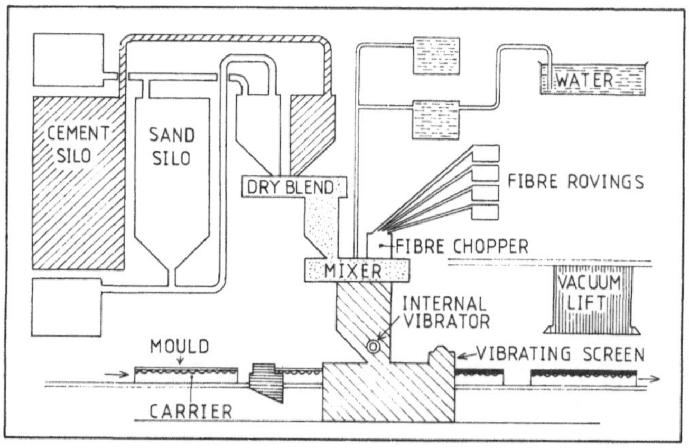

Figure 3.35 Basis of the Bancem extrusion continuous process.

Figure 3.36 Selection of products produced by the Bancem process.

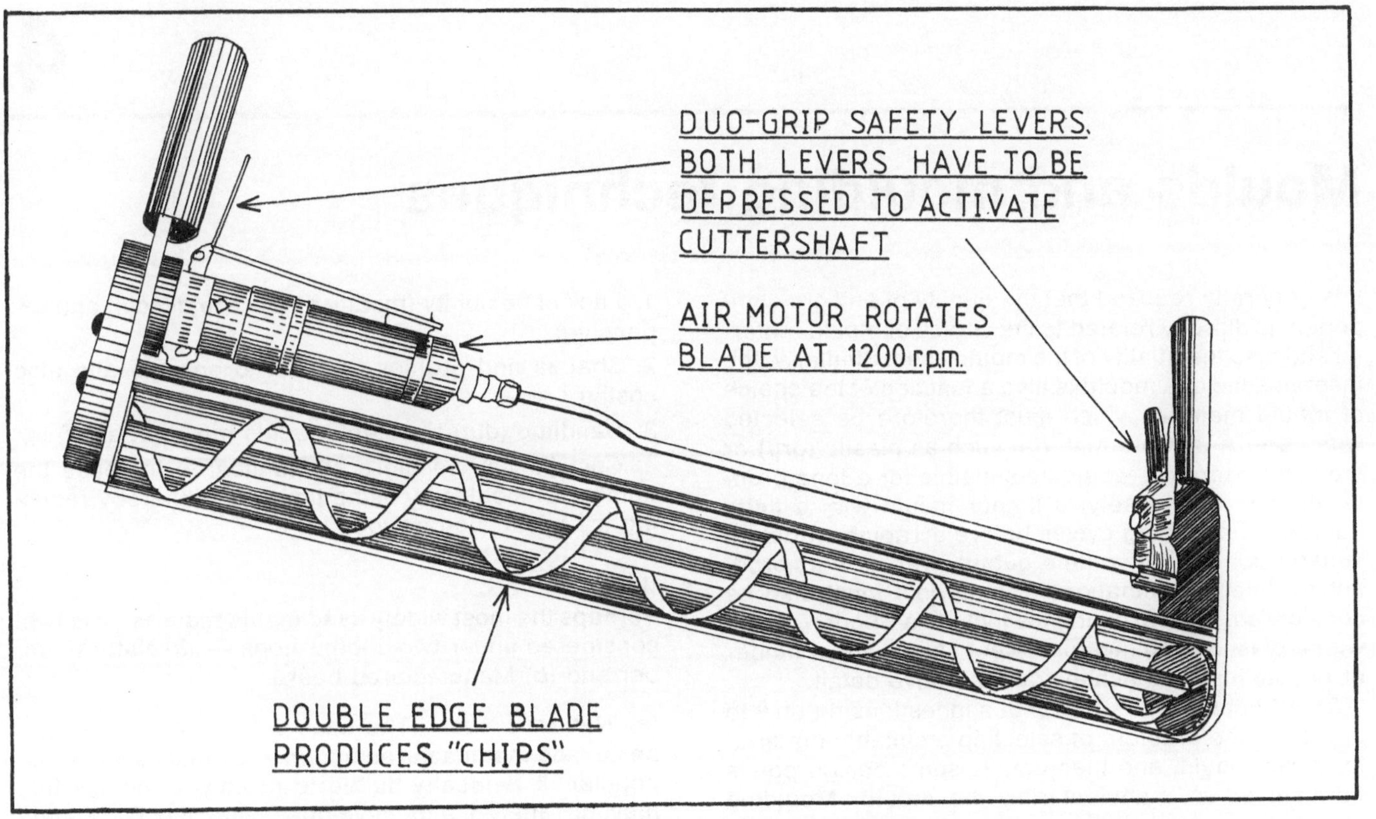

Figure 3.37 Air driven foam plane.

brought about a new generation of robot production facilities. These units are capable of working from pre-recorded tapes, usually made by an operator taking the equipment through a 'teach' sequence. Once this teach sequence has been stored on tape, the robot is able to repeat the operation at will. A whole series of such tapes can be built up, each referring to a particular sequence of operation for the production requirements of a component.

Obviously the robot facility has practical limitations and can best be employed for producing products otherwise suited to hand-spray work such as items of 'volume' requiring vertical and horizontal spraying in all three dimensions including coffer pans and street furniture.

The axes of motion provided by robots are the swing axis (rotation of the manipulator arm horizontally), the tilt axis (where the manipulator arm rotates on its vertical plane) and the extension axis (provided by a telescopic extension and retraction of the arm). A wrist action is also provided at the head of the unit giving it side to side and vertical movement.

The concentric gun has been found to be essential for this machine due to the single head action ensuring good wet-out of the composite on lay-up.

The untiring nature, constant output and repeatability of the robot unit is worthy of consideration, but it requires back-up with a roller compaction and finishing team to finalize the production process. In this light it should be seen as a pace-making facility of constant output; a placing machine that carries out a spray-up process and not as a complete facility capable of independent operation.

3.13 ANCILLARY EQUIPMENT

There are three items of ancillary equipment which should be considered:

- Compaction rollers
- Foam planes
- Turntables

3.13.1 Compaction rollers

Immediately a layer of 3 mm or so of grc has been 'laid up' it is necessary to compact this before any subsequent layers are placed on top. Compaction rollers are available for this purpose and consist of aluminium tubes approximately 50 mm in diameter with parallel flutes running radially along the tube. Surface dry fibres are 'wetted out' and the composite pressed together. This has the effect of increasing the bond of the fibre and removing air voids from the composite. The advantage obtained from roller compaction is lost if too thick a composite is built up before rolling.

3.13.2 Foam planes

The tool shown in Figure 3.37 is a convenient means of levelling off rigid foams that may be used to form sandwich construction products. It is an air driven tool and comes in a range of cutting widths from 300 mm to 600 mm.

3.13.3 Turntables

These can be invaluable when products are being manufactured that require spraying from more than one location such as coffer pans or large cladding panels. It enables the sprayer to remain stationary while the workpiece revolves. Turntables should be basic and durable to stand up to the rough treatment of production handling.

4

Moulds and moulding techniques

It is generally realized that the quality of any grc component is directly related to the choice of mould material and also the quality of the mould itself. Similarly, the inservice life of a mould is also a function of the choice of mould material which must therefore be selected with care. A durable material such as plastic (grp) or steel, for example, would be sensible for a long production run. Conversely, a timber mould with a finite number of moulding cycles before scraping would be suitable for a high volume output over a short production time. Combinations of materials must also be considered. Typical examples include a wooden mould with a plastic lined face or a steel mould with a timber or plastic insert to achieve a decorative detail.

This Chapter offers practical suggestions on how to approach the problem of selecting a suitable material for a grc mould, and then details some design points relevant when manufacturing the mould. Moulding techniques, concentrating on how the wet composite can be effectively laid into the mould and finished, are included as a final topic.

4.1 MOULDING MATERIALS

The following section lists some materials that can be most advantageously used for grc moulding. Due to limitations of space it should not, however, be seen as a definitive statement on the subject.

4.1.1 Materials suitable for the main body of the mould

Steel, wood (natural timber and manufactured board), glassfibre reinforced plastics, methyl methacrylate (perspex), concrete and grc are all materials requiring consideration as suitable for the main body of the mould.

Steel

An obvious choice, but cost considerations related to total production numbers and the rate of production output could outweigh its merits. Steel stiffeners or underbeams may be useful to reduce warp in timber moulds, especially when filled and left to cure on an irregular surface. Steel edges are beneficial to act as trowelling guides and especially for removable sides where damage can occur during removal and reassembly over the production run. Steel moulds may not produce an acceptable aesthetic finish and may require some form of liner to achieve the preferred texture and shade. If the production run requires alterations to be included, such as cladding units, then consideration for this must be made at the design stage.

The advantages of steel moulds are:

1. Long life (they easily outlive their use)
2. Accuracy (they do not have the dimension problems that occur with timber)
3. Reasonable finish.

Consideration must be given to the following:

1. Lack of flexibility (modifications are difficult and expensive)
2. Shapes and profiles (difficult to achieve and also costly)
3. Handling (durable steel moulds may be heavy)
4. Cost (it may be more economical to increase the number of moulds by using timber and thereby reducing the production period).

Wood

Perhaps the most widely used mould material, it is best considered under two subdivisions — (*a*) Natural timber and (*b*) Manufactured board.

(*a*) *Natural timber* The ease with which timber can be cut and manufactured into moulds makes its choice popular. It is ideally suited to prototype moulds that may be altered and developed with the knowledge gained from test casts. However, the life of any timber mould is firstly dependent on the design, manufacture, materials utilized, sealer treatment and complexity, and secondly affected by the handling and maintenance in service together with the ease of moulding and demoulding when in use on the shop floor.

Surface finishes obtained from timber moulds are influenced by the texture of the mould and the absorbency of the timber. These two factors can change after a succession of casts as the surface is eroded and absorbency reduced due to the 'take up' of cement particles and mould oil. It would be prudent to seal the surface of a production mould with at least a two-coat layer of polyurethane based mould sealer or line the interior of the mould with an efficient lining material.

(*b*) *Manufactured board* The most effective board for this purpose is exterior grade plywood. This is obtainable in a range of thicknesses and with a number of finishes including epoxy resins and grp. Plywood moulds can be light and durable, but require a certain amount of common sense in design. Cut edges should be covered with either metal strips or glued on hardwood edging. Corner reinforcement is recommended and, in particular, metal plate reinforcement and joiners minimize subsequent distortion.

Plywood offers great potential for the design and manufacture of grc moulds. If the material or its employment are substandard, however, then the mould will prove either difficult to maintain or produce a product which is unacceptable aesthetically and dimensionally.

Glassfibre reinforced plastics

Increasingly, grp is being used for manufacturing compound curved moulds to a fine accuracy. Such moulds would be too expensive in steel and too difficult in timber or plywood. The disadvantage with grp is that it requires a master to be produced first on which it can be laid up. Also, the manufacture of grp moulds must be considered a specialist job. However, once a mould shop has managed to acquire the knowledge for this it opens up great potential. From one master a series of moulds can be cast in a short period of time. It should, however, be noted that the mould shop requires an ambient temperature around 20 °C in order to assist the curing of the resin. In the event of the breakage of a grp mould, it is a relatively easy matter to repair and refinish the mould for replacement into the production line.

Although grp moulds have a potentially long life they should be treated with a certain amount of care, especially on demoulding. Using hammer blows to assist demoulding is not likely to ensure a long mould life and may lead to 'star cracking' of the surface gel coat. These cracks will then progressively open on subsequent casting and, if not filled with a polyester paste filler, may lead to a break up of the mould.

Methyl methacrylate (perspex)
Perhaps this is an excessively expensive material for general use, but it has certain advantages. It can be machined to size and tapped to accept screw threads. Consequently, it is ideal for small moulds that would have a long production run. Another advantage is that it can be heated and shaped over a former.

Concrete and grc
Both concrete and grc are ideal durable moulding materials. Concrete, however, produces a heavy mould that is best considered as static rather than movable around the factory. Grc can be used to produce fine textured and high relief moulds, but it must be noted that no give is experienced on demoulding and so any draw on the mould must be adequate to ensure this is easy. Concrete and grc may form the basic element of the mould with perhaps a plastic, wooden, or steel edging, that is removable to assist demoulding.

Both concrete and grc moulds require curing before use and so consideration of this must be made in the production programme.

4.1.2 Materials suitable as components of composite moulds

Often moulds can be best produced by a composite construction. The main body of the mould would be produced in one of the preceding materials, i.e. steel, wood or grp, with a liner of some other material. This may be a flexible rubber, a plastic or possibly even a thin wooden board such as hardboard. The purpose in using this method may be to reduce the cost of the mould by keeping the expensive liner (in the case of flexible vinyl rubbers) to a minimum volume of material, and also to assist in the demouldng of grc.

The following are mould lining materials requiring consideration.

Hardboard
Hardboard is used to produce an absorbent, low cost, dispensable lining for an aesthetic purpose.

Melamine laminates (formica)
This material forms a durable, resistant lining to timber moulds. It is usually bonded to the wood with water resistant glues, and care must be taken with edges to prevent water ingress under the laminate. A corner fillet of hardwood, which gives protection to the laminate, is beneficial in reducing the risk of debonding and also forms an arris free corner.

Vinyl rubber
There are two systems of castable rubber moulds, first a cold curing two-pack system and secondly a meltable system. The latter is more expensive but, when purchased with the melting equipment, has the advantage of being cycled after each contract to produce other moulds. Perhaps the most common use for this type of mould is in the production of a profiled mould surface with overhangs and such like. The rubber is held in shape by an outer mould case of wood, steel or grp and, when the grc product is demoulded, the rubber can easily be peeled from the component. A range of hardnesses is available with varying degrees of flexibility to suit particular moulding characteristics. Usually these rubber moulds require no mould oil to effect a release.

Neoprene rubber
This material is available in a range of textures that can be exploited to produce a profiled finish to grc. Thin sheets are best stuck down to the containing mould to prevent warping or distortion but thick sheets require only a random tacking down with glue pads.

Mineral oil must not be used as the release agent since it causes the rubber to deteriorate. The mould oil should be a vegetable or animal oil. Both castor oil and lanolin have been successful. It should also be noted that fine profiled ribs or patterns require longer curing times before demoulding in order to gain sufficient strength to prevent breakage and spalling. At least 30 uses should be possible from a rubber lining. Mould oil must be carefully applied to textured neoprene linings to prevent localized ponding within the mould texture that leads to porous, weak and possibly stained surfaces on the grc. Profiled or textured surfaces taken from a shallow neoprene lining may well be produced without any mould release agent.

Polypropylene
This plastic is an ideal edging material for timber or steel moulds. It can easily be cut by hand with a hacksaw or a handsaw. It is flexible, so can be made to follow contoured moulds, but is tough enough to withstand the trowelling and finishing of the grc without damage to its surface. Any stepped or special edging should be achieved by a double strip to form the required contour. Cleaning of the polypropylene edge is a simple matter due to the lack of bond achieved between the grc and the polypropylene. Fixing of the mould edge should be obtained by bolting down to the mould base through holes drilled in the polypropylene edge strips. It may be possible to fix down the edges as a permanent mould detail without having to make them removable for demoulding.

Linoleum
This cheap material can sometimes form a simple yet adequate lining to achieve a radius at mould edges.

Rigid foam plastics
Foam plastics, such as polyurethane and polystyrene can provide a cheap, highly textured lining suitable as

a disposable mould lining. This would probably be bought preformed and used to produce a decorative panel or surface to a cladding panel or planter.

4.2 MOULD DESIGN

The design of grc moulds should take the following points into consideration:

1. All moulds must be robust enough to take the day-to-day wear common in production. Naturally, the number of castings made from the mould would to some extent dictate the material choice for the mould. Deflection in the mould due to the cast-in grc and to the mould weight itself must be considered.

2. Ideally, all corners should be to a radius of 10 mm (minimum 5 mm) or have fillets of 45°. Draw angles of 5° are the optimum.

3. Cast-in jacking points help demoulding and increase the mould life. In-built air connections are also significant in this respect.

4. Hinged or detachable mould sides aid withdrawal of the grc product (Figure 4.1). Split moulds have the same advantages for complex edges (Figure 4.2).

Figure 4.1 Hinged or detachable mould sides to aid product withdrawal

Figure 4.2 Split moulds for complex sides.

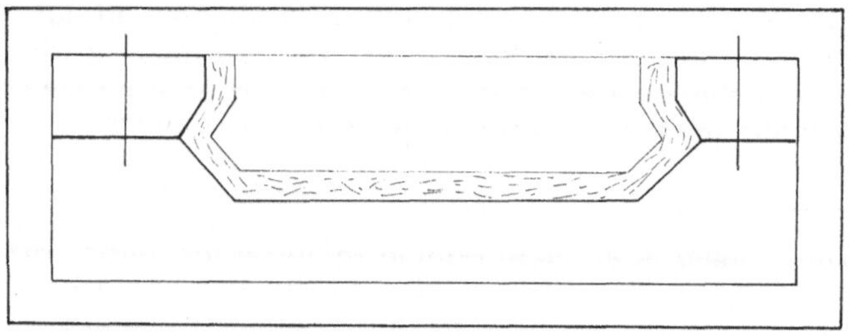

5. The spray gun must be able to gain access to all parts of the mould. Hand placing into inaccessible mould features is a poor design feature and will produce a weak, porous composite, instead of the required durable, well compacted, high quality grc needed for service conditions. Figure 4.3 indicates the relationship of width to depth in a mould highlighting the access potential. If intersection falls on a hatched area then spray access is possible. The dark zone indicates marginal accessibility.

6. Spray production gives only one mould face unless bonded panels are produced. Non-moulded faces will have a textured or trowelled finish that is only as good as the operator's skill. Overworking exposes fibres at the surfaces.

7. Surface ribs and channels should be wide enough to allow roller access for adequate compaction. For example, a 12 mm groove width should not be more than 5 mm in depth. Fine detail will require extra compaction with a stiff brush prior to roller compaction. In this case the arris will be void of fibres and highly likely to be spoilt on demoulding and handling due to spalling.

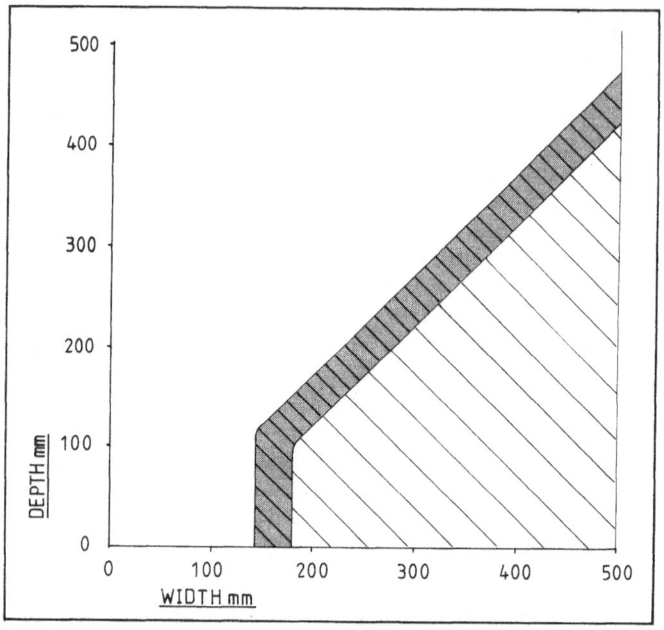

Figure 4.3 Relationship of mould width to depth where access for spraying could be difficult.

8. Small details in the mould will take the worst battering. If made from timber they may need frequent replacement. Stainless steel or plastics (polypropylene, etc.) are more suitable. This is shown in the drip detail (Figure 4.4) where a hinged side former containing a stainless steel section forms the drip detail that can be pressed into the trowelled back of the grc.

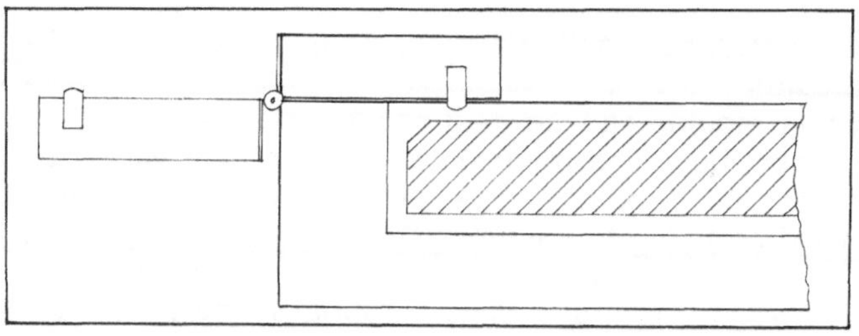

Figure 4.4 Hinged pressure plate to form feature in trowelled face.

It should be noted that localized thickening of the single skin of panels will lead to ghosting at the surface and may mar the aesthetic value. Ribs produced by styropor concrete or foam plastics can be used to stiffen the panels with less chance of surface ghosting.

4.3 MOULDING TECHNIQUES

The following sections detail in easy steps, and with illustrations, suitable moulding techniques.

4.3.1. Edges

Flat products

A. Return edge is attached to the mould with depth equal to the design thickness and of width at least 50 mm.

B. The grc is sprayed into the mould, tucking the material into the corner with special purpose tool. Final layer not sprayed yet.

C. Overspray folded over, back onto panel surface.

D. Final layer is sprayed, rolled and trowelled to achieve design thickness.

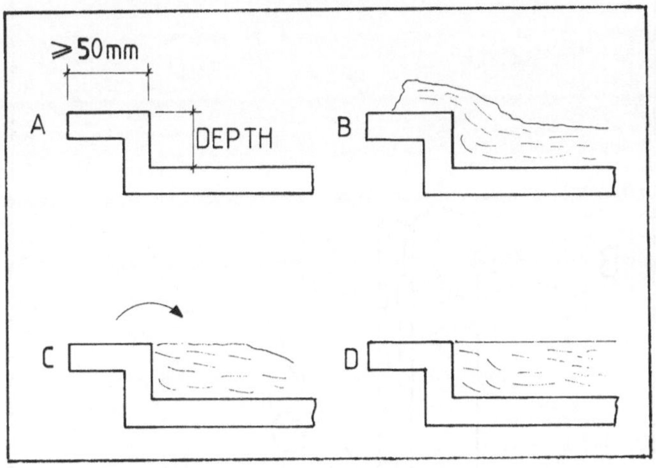

Upstand edges

Trowelled edge

A. Mould edge as shown.

B. Spray in layers as in above.

C. Overspray folded back onto panel.

D. Final layer sprayed to achieve design thickness. Rolled and trowelled to give a neat flag edge.

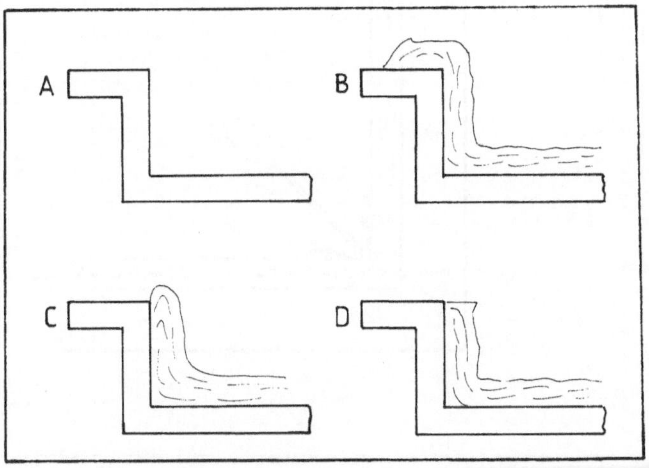

Moulded edge

A. Mould edge as shown.

B. Sprayed in layers up to final layer. Material tucked into corner with special purpose tool.

C. Overspray tucked back onto panel. Final layer sprayed to design thickness.

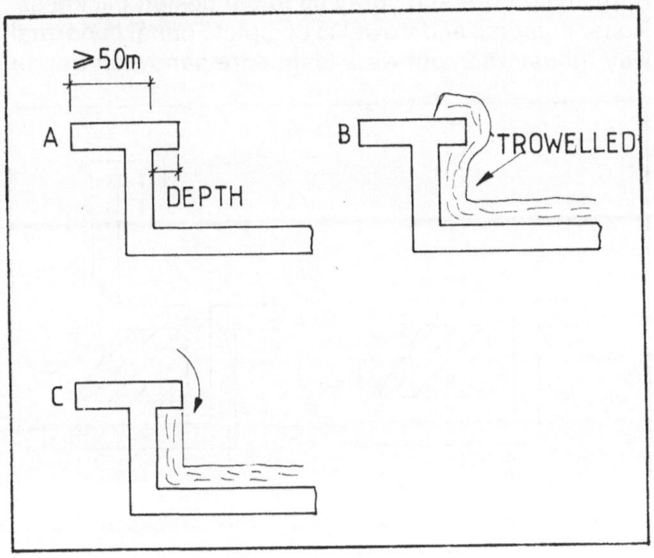

Ribbed edge

A. Sprayed in layers, roller compaction and checked to ensure correct thickness on side-wall and base.

B. Precut foam former (as shown) placed in mould and checked with gauge to ensure that the former is D mm below mould top.

C. Spray over former to meet last spray. Turn over overspray from mould edge and roller compact.

D. Spray to correct thickness. Pay attention to thickness at a and b and ensure that the overlap at b is sufficient.

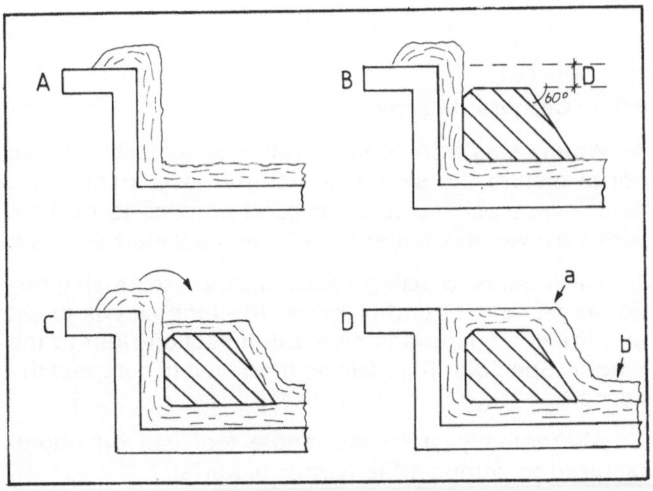

Rising shutter edge

A. Spray base of shutter as normal. Then lay foam core as described for ribbed edge method above.

B. Locate rising shutter and spray up vertical face and over mould edge.

C. Fold back overspray; place rib foam core in place. Spray over core and spray up to the design thickness. Roller compact and trowel to complete panel in normal way. (Illustration shows a foam core sandwich panel)

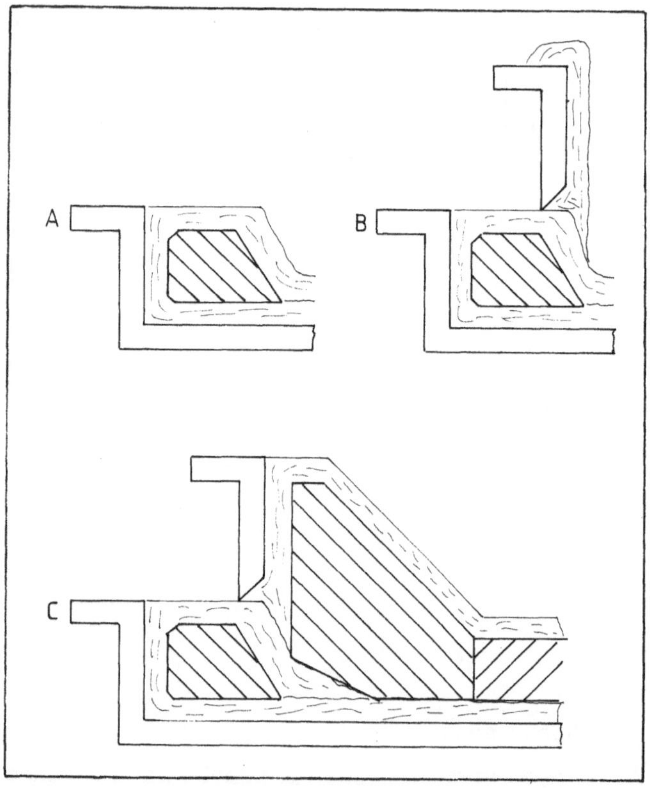

4.3.2. Corners (internal)

A. When using a compaction roller care must be taken not to disturb the side wall material (see arrow). If a disturbance occurs, it is not good practice to replace with a trowel. It is better to respray the disturbed area.

B. This method of using a float or trowel to finish internal edges is not good practice. The tendency is to cut into the composite and give a thin section right in the corner. Special rollers can be produced to compact the corners.

C. Alternatively, a special angle tool can be manufactured to compact the corner material.

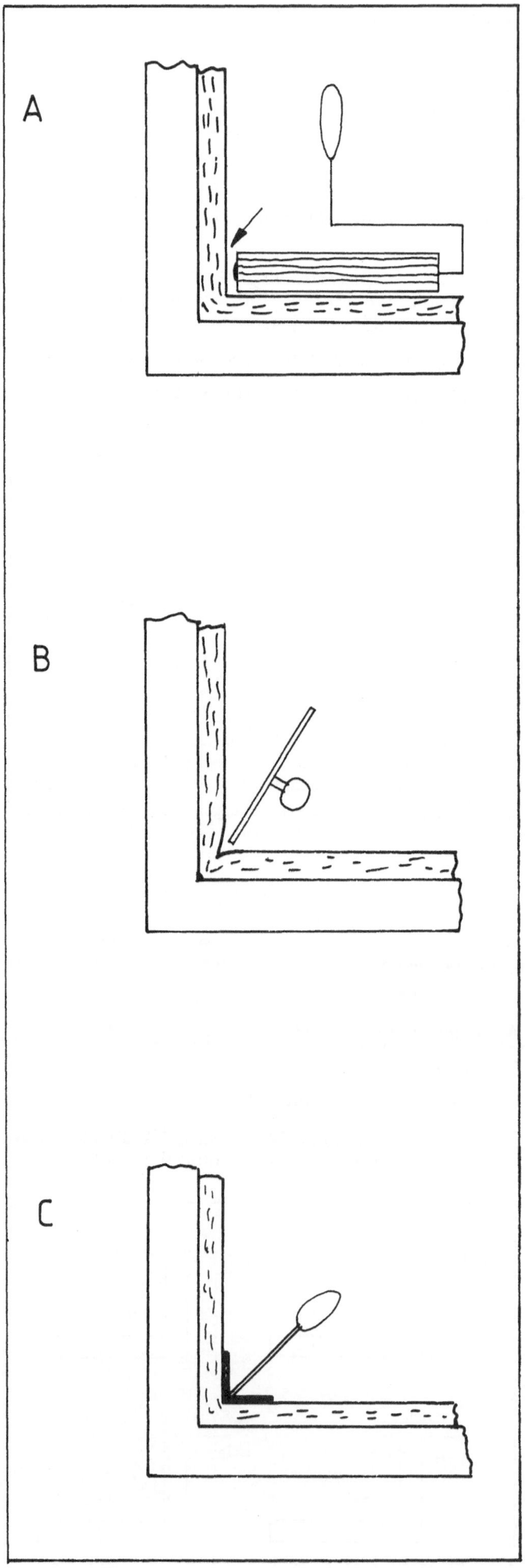

4.3.3. Corners (external)

A. If support is not given to the edge of the grc (as shown) the material will spread and squeeze away from the mould leaving a pocket or void down the edge. Thinning will also occur.

B. The above situation can be avoided by supporting the side of the sprayed material with a trowel, while roller compacting or trowelling the top edge.

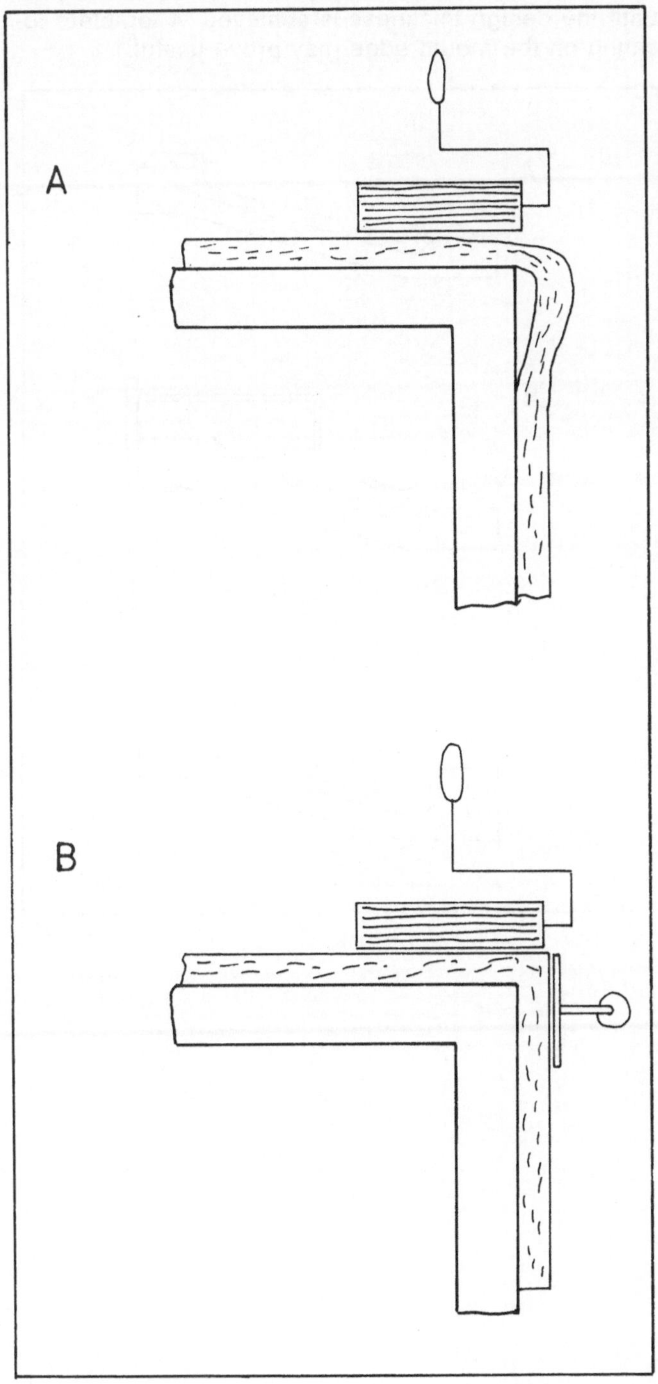

4.3.4. Sandwich panels

Organic foam core

A. Spray and compact mould face and sides as normal, checking thickness.

B. Precut foam blocks, as shown, to fit loosely together and are coated with mortar and binding agent (e.g. Acronal). These are placed in the mould with chamfered blocks at the edges. Care is required not to disturb the vertical grc edges.

C. Check top surface of the foam core is at correct height as shown. Spray 3 mm layer of grc, then fold over overspray from edge and roller compact.

D. Complete spraying top surface of foam and roller compact. Finish with trowel and straight edge to check level.

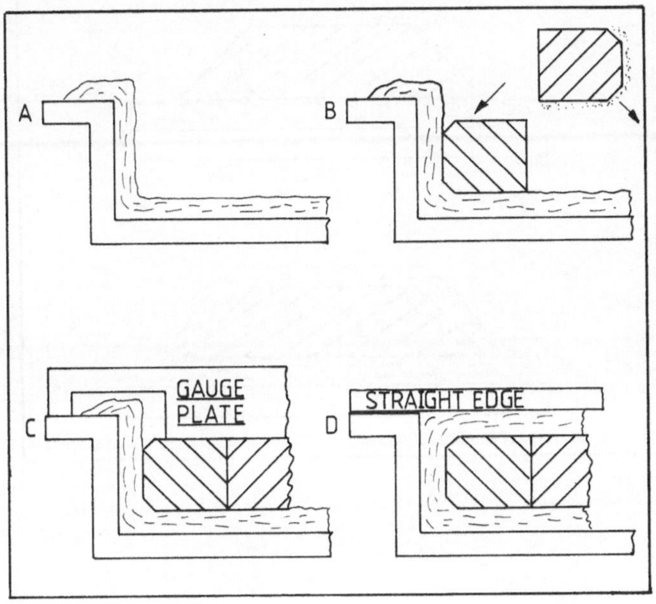

Styrofoam concrete core

A. Spray and compact mould face and sides as normal, checking thickness.

B. Fill mould with styropor concrete mix. Tamp to remove air pockets. Use gauge plate to check depth. Spray in 3 mm layer of grc and fold over overspray from the mould edge and roller compact.

C. Spray over with grc in layers and roller compact to achieve the required thickness.

D. Ensure that a grid of 6 mm air holes is present in the back face since the entrained air in the styropor concrete will expand and bow the back if pressure is allowed to build up.

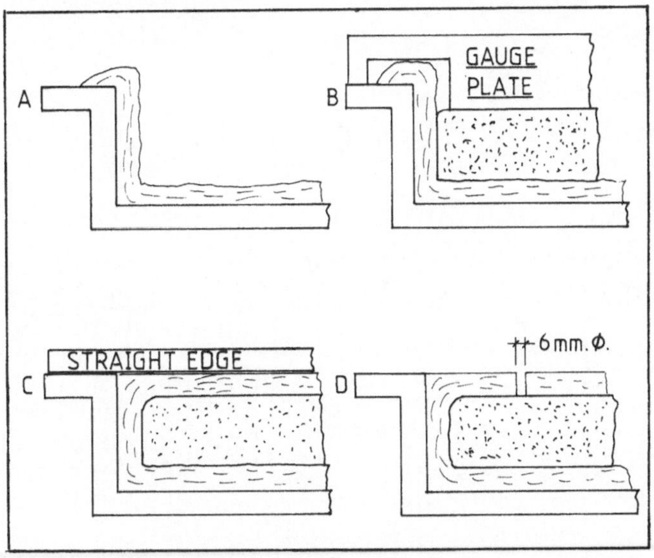

4.3.5. Ribs

Organic foam core

A. Spray front face in layers and roller compact. Check thickness, then place foam core in position.

B. Spray over sides and back of former in layers. Check thickness especially at edges of the former and at the top/side junction.

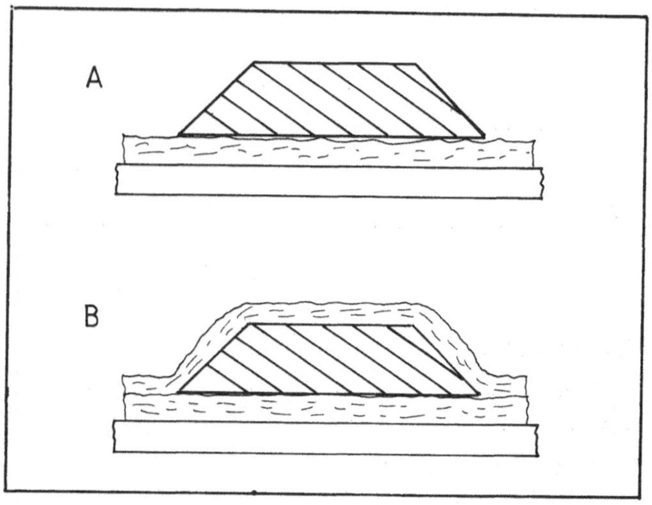

Grc core

A. As above but using a 3 → 5 mm grc former. This cannot be counted as part of the rib thickness for design purposes.

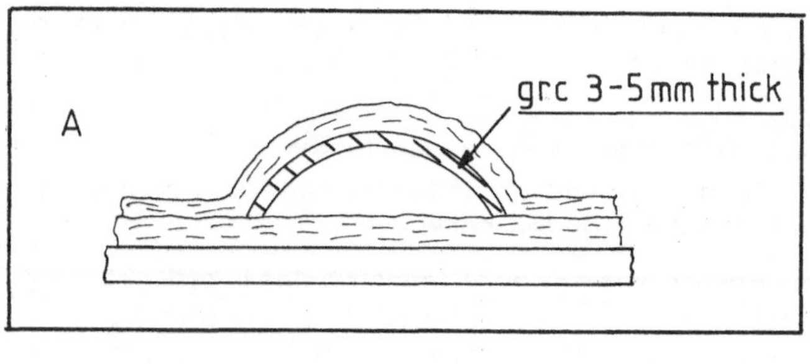

Grc diaphragm

A. Spray and compact in the normal way. Allow at least 50—100 mm overspray at the edges.

B. Locate pre-sprayed grc diaphragm using jigs or fixings to align and position. Spray 3 mm on the diaphragm and turn in the overspray and compact with a roller.

C. Spray on the back, in layers, and roller compact until the design thickness is achieved. A template locating on the mould edge may prove useful.

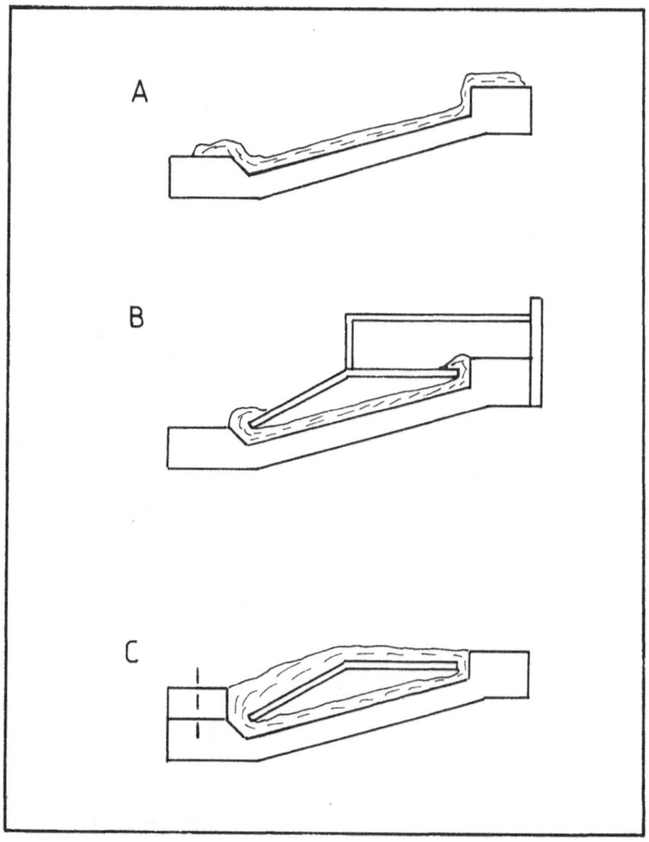

4.3.6. Fixings

Cast-in sockets

Single skin panels

A. Spray base of the panel. Build up thickness under location of the fixing point.

B. Suspend fixing on locating jig and press into sprayed material.

C. Complete spraying to correct depth compacting sprayed material around fixing.

Note: The material around the fixing should be well compacted. Also the thickness D must be at least the design thickness.

Sandwich panels

Build up thickness under the fixing point in the mould as in A above. Locate fixing. Spray over fixing and onto base. Place core material. Spray carefully around socket (with compaction). Turn back overspray and finish spraying top layer.
(d = bolt diameter)

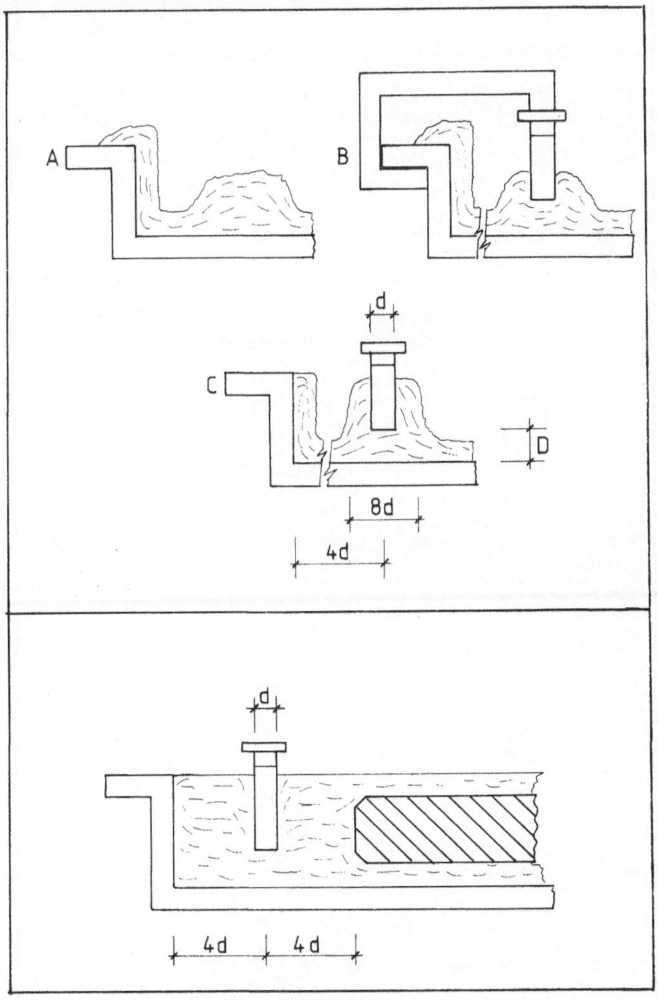

Cast-in washers

Spray and compact up to shoulder on locating peg. Place perforated washer on peg and spray compact to design thickness.
Top of peg acts as thickness gauge, if required.

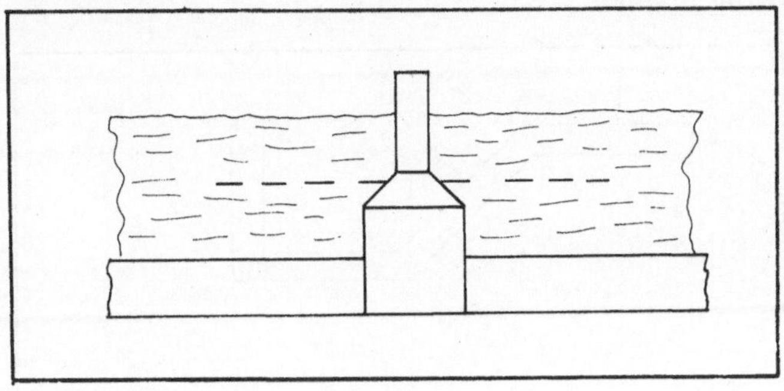

Cast-in sleeves

A. Proceed as with cast-in socket fixings but with the sleeve in position before start of spraying.

B. Spray around sleeve (with compaction) filling hole. Turn back overspray and finish spraying top layer.
(d = sleeve diameter)

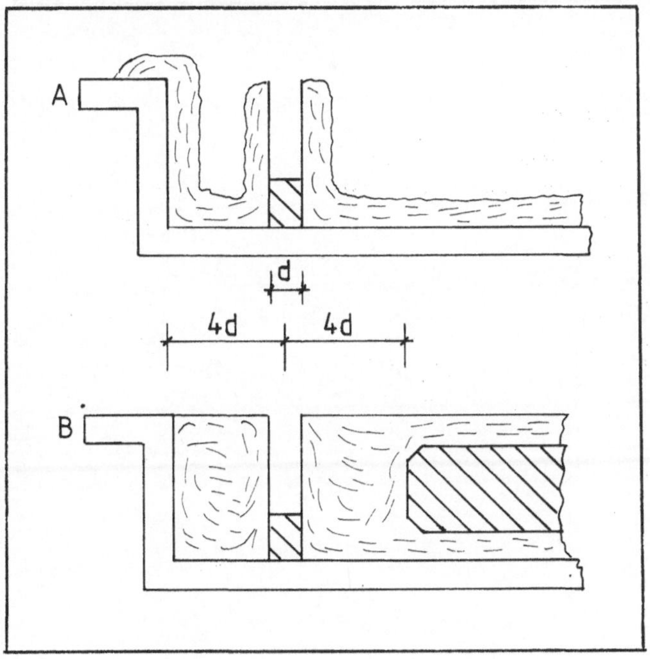

4.3.7. Hinged moulds

A. With mould edge down, spray up in layers as in the normal manner, and compact.

B. Fold up mould side, clamp in position, roll or tamp with trowel in hinge area, finish top edge of return in usual manner.

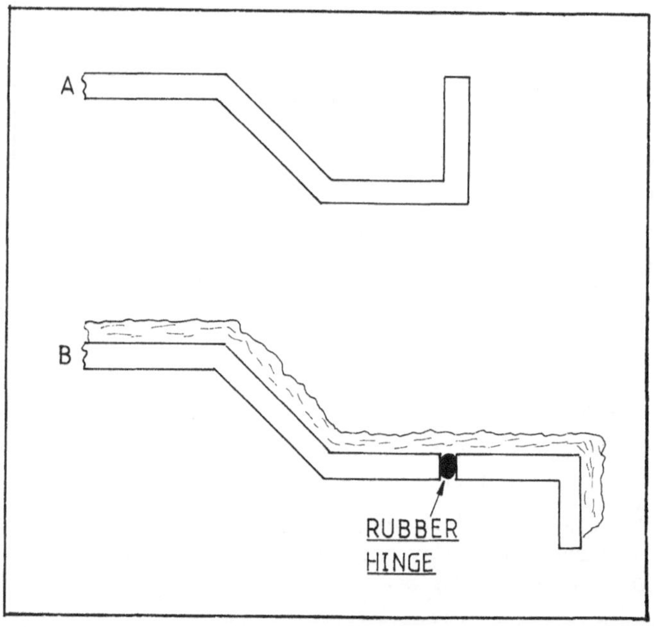

4.3.8. Textures and profiles

A. Special tools may be required to suit the profile of certain designs. These will ensure that the grc is well compacted into the profile.

B. Certain light profiles and textures may require the application of a stiff brush before roller compaction.

C. Sharp profiles may require a wire or rod tool to ensure compaction is achieved in the corner of the profile.

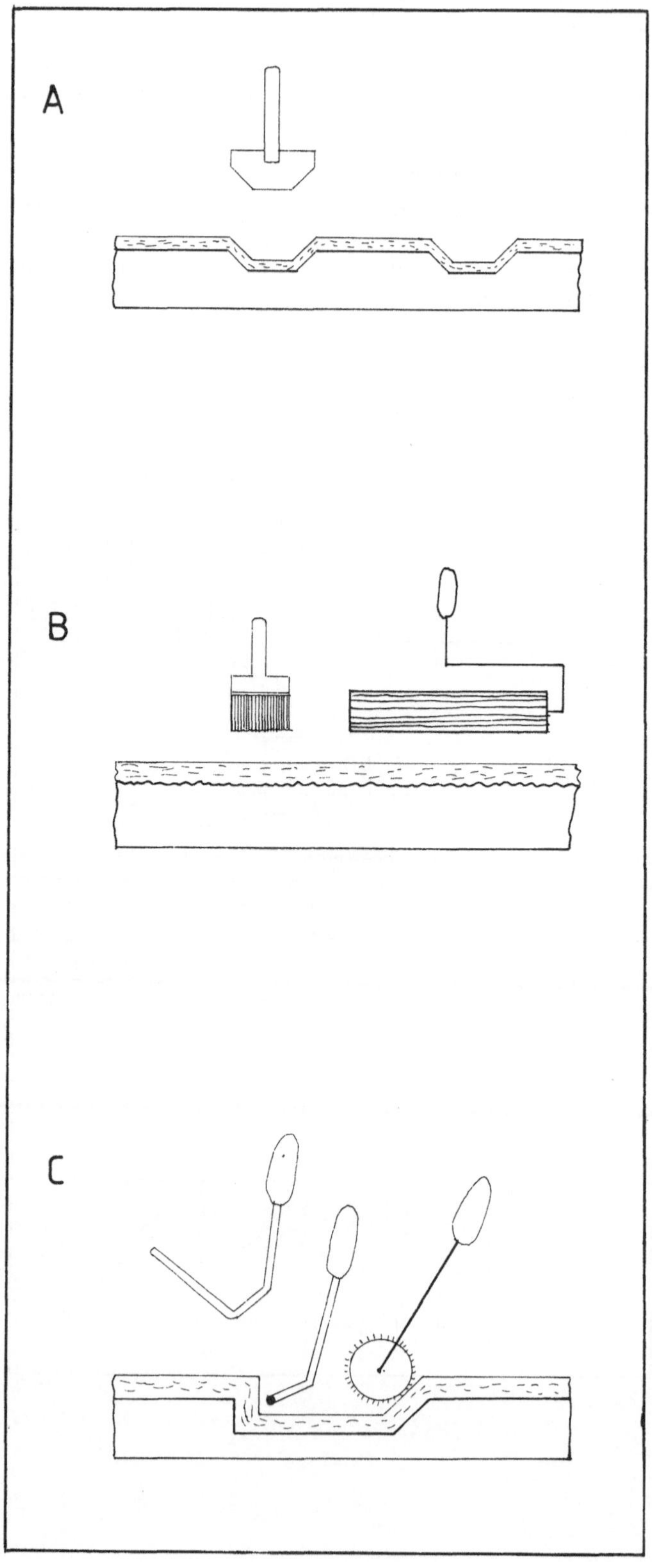

5

Constituent materials

Fibre, cements, sand/fillers, pigments, admixtures and lightweight core materials are all considered as constituents of grc. Besides examining these conventional substances additional materials worthy of further investigation are also discussed within this Chapter.

5.1 FIBRE

Suitable fibre additions and blends are detailed in this first section.

5.1.1 Glassfibre

A variety of types of glassfibre have been incorporated into cementitious matrices. These can be divided into two main categories:

- Alkali resistant
- Non-alkali resistant.

Alkali resistant glassfibre

Up until 10 years ago the only glassfibre available was the non-alkali type, the kind used in glassfibre reinforced plastic manufacture. Investigations carried out by the Building Research Establishment discovered that greatly improved resistance to alkali attack was possible by including zirconia sand in the glass formulation. A similar formulation had been discovered some few years previously in the USA for manufacturing blood sample storage phials.

The successful Na_2O–CaO–ZrO_2SiO_2 system was patented by the N.R.D.C. and eventually developed into a viable process for manufacturing into glassfibre. This fibre has the brand name Cem-FIL and is produced by Pilkington Brothers of St Helens, Lancashire.

Other formulations have been developed and are marketed in the USA, Europe and Japan, but to-date the Pilkington Cem-FIL product has had the greatest use. Two forms are currently available (see Figure 5.1):

(a) Rovings

(b) Chopped strands.

Figure 5.1 Illustrations of roving and chopped strand fibres (as well as sand and cement).

(a) Rovings are continuous strands of glassfibre grouped together in a parallel bundle and wound to form a square-ended package referred to as a cheese. This usually weighs around 21 kg.

The individual fibres are sprayed during the manufacturing process with a size, an organic water-based emulsion formulated to bind the individual filaments together in a 'strand'. This coating also protects the filaments against abrasion and fracture during processing and achieves a suitable bond between the roving and the matrix. The individual filaments are of the order of 10 to 20 microns diameter.

(b) Chopped strands consist of rovings chopped into uniform lengths of between 6 mm and 50 mm. They are purchased by specifying the cut length. Since chopped strand fibres are used in the pre-mix process, greater strand integrity is required. This is obtained by modifying the coating applied to bind the filaments. A stiffer fibre is produced that has greater integrity and can resist the abrasion caused by the mixing process.

Both rovings and chopped strands require careful treatment avoiding extremes of temperature and humidity. Damp and dust should be kept from contaminating the fibre, and roving boxes must be sensibly handled to prevent crushing and damaging the cheese. Ideal storage conditions are 10–25 °C at 40–65% RH.

In the past, a third form of alkali resistant glassfibre has been available. This was the non-woven mesh called crenette. While very cost effective for press moulding, it has been little used and is now unobtainable.

Non-alkali resistant glassfibre

Before alkali resistant glassfibre was readily available, 'E' glass and 'A' glassfibres were used in roving and in chopped strand mat forms. The cement employed for this was aluminous cement, having the advantage of a far lower rate of attack on the fibre than ordinary Portland cement. The 'E' glass or borosilicate glass was more widely used and, in this case, sizes of a paraffin emulsion base have reputedly been successful in binding the filaments together. When aluminous cement is specified because sulphate attack or chemical attack is suspected, as in permanent formwork for pipes or sewers and such like, it may be cost effective to consider 'E' glass as the reinforcement.

5.1.2 Other fibres

As the development of fibrous cement composites continues, more economic fibres and fibre 'blends' have evolved. Nylon and polypropylene fibres have, for example, excellent impact properties in cement matrices but offer little tensile strength due to their low modulus and poor bonding characteristics. Blending nylon or polypropylene as a premix into the matrix and then manufacturing by spray-up using gun chopped rovings of glassfibre can produce composites of good impact and tensile strength. The fibre addition levels may be in the order of 0.5–1% of nylon or polypropylene of 12 mm fibre length and 4–4.5% of glassfibre (both percentages by weight).

One noteworthy advantage of formulating nylon or polypropylene into the composite to handle the impact requirements is that the impact value obtained remains with ageing. This contrasts with glassfibre composites which, on their own, suffer substantial reductions in impact properties after a few years weathering.

Perhaps the most exciting development in blending glassfibre with other fibres is the recently developed high modulus nylon — Kevlar 49. This fibre, when introduced into the tensile zone of grc composites, considerably improves the stiffness of the product, increases the LOP value, maintains the improvement with age, and also reduces the shrinkage and creep rates. It is, however, four times the cost of alkali resistant glassfibre, and has reduced properties at elevated temperatures compared with inorganic fibres. Nevertheless, for permanent formwork units, pipes and so on it has great promise.

5.2 CEMENTS

Most general applications of grc production employ the ordinary types of cement that are readily available namely:

- Ordinary Portland cement
- White Portland cement
- Rapid hardening Portland cement
- Sulphate resisting Portland cement
- High alumina cement
- Ultra high early strength cement.

Further advice on the use of these should be sought from the manufacturers, but the following comments are offered as a guide to assist in a general consideration of their respective merits.

General production should employ rapid hardening Portland cement. The extra hydration over ordinary Portland cement obtained at demoulding (usually 18 hours after casting) is sufficient to produce a marked benefit in strength, with only a marginal cost increase.

If white composites are required, then naturally white ordinary Portland cement has to be used, but it may be possible to apply white cement only for the mist coat (with a polymer admixture), and back this up with rapid hardening cement mortar for the bulk of the composite. The cost differential alone makes this worthy of consideration and development.

Situations that would require sulphate resisting cement may not warrant this choice with grc formulations, but it may be prudent to include pulverized-fuel ash in the matrix instead of sand.

If circumstances necessitate high alumina cement in the formulation, then consideration should be given to the possibility of using 'E' glass as reinforcement.

When rapid demoulding is required in production, regulated set cement may be considered. This cement can produce a set 5 minutes after mixing. Consequently, purpose-built mixing and dispensing equipment would be required. This rapid setting gives a strength that may allow demoulding in just 15 minutes after mixing the matrix. The rapid early strength build-up is maintained at a constant level until the normal hydration products produce strength gains that exceed this early strength level. When considering regulated set cement, the extra cost may offset the benefits unless large production outputs are envisaged.

5.3 SAND/FILLERS

5.3.1 Sand

Sand is introduced to reduce shrinkage of the composite. A 25% cement replacement will provide a reduction of 50% (see Figure 5.2). Shrinkage is undesirable if restraint occurs via fixings or product shape. Sufficient stress can build up resulting in cracking when the shrinkage exceeds the tensile strength of the section in question.

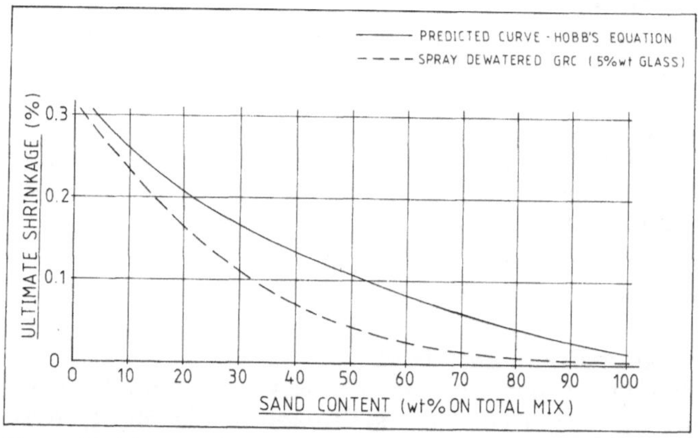

Figure 5.2 Ultimate shrinkage as related to sand content in the matrix.

Dry sand is preferred since an accurate check can be kept on the water incorporated in the mix. The sand must be free of large particles which would block the spray nozzle. Also, clay lumps and silt in general are undesirable as they can increase the water demand. For these reasons, washed, graded and dried silica sand which is free from soluble salts, clay particles and organic matter is strongly recommended. Table 5.1 details a viable specification.

Table 5.1 Specification for sand.

Composition	Silica	96–98%
	Moisture	2%
	Soluble salts (sulphates, chlorides and alkalies)	1%
	Loss of ignition	0.5%
Grading	Zone 4 (maximum particle size 1 mm)	
	Fine fraction < 10% passing 150 μm sieve.	

Suitable sands are available in bagged or tanker form throughout the UK usually specified as grade 50.

5.3.2 Pulverized-fuel ash (pfa)

Pfa is a widely available mineral by-product of pulverized coal fired power stations. However, there is a wide variety of types and fineness of ash available due to the variations in types of coal, boiler efficiency and ash collection plant. The C.E.G.B. have therefore made great efforts to ensure that the output of pfa is relatively consistent and conforming to the specification given in Table 5.2.

Table 5.2 Specification for pulverized-fuel ash.

Constituent	Max. %	Min. %	Typical %
Silica (as SiO_2)	58	38	48
Alumina (as Al_2O_3)	40	20	26
Iron oxide (Fe_2O_3)	16	6	10
Calcium (CaO)	10	2	4
Magnesium (MgO)	3.5	1	2
Sulphate (SO_3)	2.5	0.5	1.2
Alkalies (Na_2O,K_2O)	5.5	2.2	4.5

Handling is similar to that for cement. The material is available dry in bags or tankers and may be stored in silos.

It should be noted that:

1. Bulk density of dry pfa is around 900 kg/m³
2. The material should not be stored too long in the silo because of the danger of packing and consequent difficulty in discharging
3. Due to its lightness, pfa tends to 'fly' more easily than cement. Drops into the mixer should be kept to a minimum and ideally an enclosure provided to prevent losses.

Pfa is most readily specified by reference to particle size, grading, carbon content and specific surface area. It is also marketed by a company called Pozzolanic Ltd. They offer a material surpassing the requirements of BS 3892 pulverized-fuel ash for use in concrete. The specification for pozzolanic pfa is given in Table 5.3.

Table 5.3 Specification for pozzolanic pulverized-fuel ash.

Specific surface	Normally greater than 325 m²/kg satisfying BS 3892 Zone B requirement	
	Pozzolanic material	BS 3892
Magnesia (MgO)	2% by wt	4% by wt
Sulphate (SO_3)	1% by wt	2.5% by wt
Loss on ignition	4% by wt	7% by wt
Moisture content loss to 110 °C	0.5% by wt	1.5% by wt
Carbon (C)	3% by wt	—
Grading	BS sieve size	
	150 microns	98% passing
	45 microns	90% passing

5.4 PIGMENTS

Pigments used in grc are standard concreting pigments viz. powders or liquids. It should be noted, however, that the addition rate to the matrix is quite considerable due to the high level of cement content in the formulation. To achieve any degree of toning may require a considerable addition of pigment. Lighter shades can be achieved more simply by using a combination of white oPc and pigment.

Before pigmented products are produced it should be taken into account that grc is highly prone to lime bloom formation. This can easily spoil an architectural panel or any product having an aesthetic function. The darker shades tend to suffer more in this respect. Naturally low water/cement ratios that produce a less permeable matrix reduce the problem, as well as incorporating integral waterproofers, but the disfigurement can be random and unpredictable. The least troublesome matrix is white oPc, where any lime bloom produced on the surface is hardly noticeable.

5.5 ADMIXTURES

Given below are the admixtures most likely to be called upon in grc production together with some reference to their particular function:

- Dispersing aids
- Thixotropic admixtures
- Workability aids
- Air-entraining admixtures
- Set accelerators
- Flexibility and permeability improvers
- Retarders

5.5.1 Dispersing aids

These usually take the form of non-ionic water soluble cellulose ethers namely methyl cellulose and hydroxyethyl cellulose. The ionic sodium carboxymethyl cellulose grades are not preferred since the Ca^{++} ions present in the cement system would react with the SCMC to form a water insoluble salt.

The action experienced is one of a deflocculating effect, breaking down the matrix components into the basic particles giving uniform dispersion. This can be achieved by a good high shear mechanical activation, in which case the methyl cellulose or hydroxyethyl cellulose should be added immediately after processing at high shear in order to maintain the resulting improvement during pumping and spray-up. The benefit in workability obtained can sometimes lead to a lowering of water content. This latter point can be used to counteract the reduction in early and final strength values experienced with formulations using cellulose ethers.

5.5.2 Thixotropic admixtures

Where vertical surfaces have to be sprayed and slumping is experienced with non-modified formulations, additions of polyethylene oxide have been found to induce a thixotropic effect into the composite. This may prove a successful countermeasure with the added bonus that polyethylene oxide has no retarding or strength reducing affect. The materials should be dispensed into the gauging water before adding to the mixer.

5.5.3 Workability aids

These achieve improvements in the workability of the matrix by flocculation or dispersion. Workability can also be maintained while reducing the water/cement ratio. There are two main materials:

- Ordinary plasticisers
- Super plasticisers.

Ordinary plasticisers
These divide into two types:

(a) Lignosulphonate salts — usually calcium or sodium lignosulphonates — that may retard in certain cases. Air-entrainment can also occur.

(b) Polyhydroxy carboxylic acid derivatives — many forms are available, some causing retardation.

Super plasticisers
These are recently introduced admixtures that drastically improve the workability of mortar and concrete mixes. There are two main classifications:

- Naphthalene sulphonate formaldehyde condensates
- Sulphonated melamine formaldehyde condensates.

The action of both is similar but better performance is achieved by adding the admixture after preliminary mixing. It must be noted that the improved workability may only last for some 30 to 40 minutes.

5.5.4 Air-entraining admixtures
The purpose of including an air-entraining agent into a grc formulation is to produce discrete uniform bubbles (0.05–1.5 mm diameter) at 0.1 mm spacing or thereabouts and stabilize this system. The bubbles are introduced by the mixing action and become entrained by the admixture's surfactant, anionic action.

The normal consequence of air entrainment is a reduction in strength, improved freeze/thaw resistance and reduced vapour permeability while the resistance to water penetration is improved. Grc however benefits in its increased fire resistance by the fact that water vapour can more easily escape, reducing the risk of sudden explosive spalling experienced with well-compacted material.

Materials used are vinsol resins, synthetic detergents (lignosulphonates), sulphonated hydrocarbons and naphthalene sulphonate.

5.5.5 Set accelerators
In order that rapid mould turn round can be accomplished, set accelerators may well enable certain products to be demoulded relatively quickly. Trials are necessary before production runs can be confidently undertaken.

The most predictable accelerator of both setting time and hardening rate is calcium chloride, but it has notorious side effects. Other materials that can be tried are alkali carbonates, like carbon dioxide which, at the rate of 1–2%, reduces setting time to a few minutes. Ammonium carbonate, aluminium chloride and magnesium chloride also have a considerable accelerating action. Finally and notably, high alumina cement at addition rates of up to 20% of the oPc content, progressively decreases setting time.

5.5.6 Flexibility and permeability improvers
Grc matures to become a brittle low strain capacity material after only a few years in a natural environment (air curing). This reduction in flexibility, together with the high shrinkages are brought about by the summation of hydration, drying and carbonation shrinkage. It can inevitably lead to cracking and crazing, if the consequential stresses induced by the summation of the shrinkage mechanisms are not relieved by creep. The addition of a liquid polymer, usually an acrylic or styrene-butadiene copolymer has been found to prevent this crazing and produce a more 'tolerant' material.

Other benefits are that permeability is more consistent and tensile and bending strengths are improved. In addition the glass has been sufficiently protected to prevent the ingress of hydration products into the fibre bundles. This enables 'E' glass to be used instead of the AR alkali resistant material.

5.5.7 Retarders
When a complicated product is being produced that may require the inclusion of a lightweight filling material for insulation and so forth, together with fixing and other accessories, it may be beneficial to add a set retarder to the formulation. This ensures that the first layers of grc placed into the mould have not passed the initial setting phase prior to being turned over at the edge and trowelled into the final packing coat. A good bond between the sides and the back sheet of the product will, therefore, be achieved.

Premature setting during manufacture can easily occur in the summer months and additions of set retarders may therefore be required. In hot climates their use is obviously more general. The materials usually employed are hydroxycarboxylic acid or polyhydroxy compounds.

5.6 LIGHTWEIGHT CORE MATERIALS
These divide, basically, into cement bound materials, board/sheet material and spray-in foams.

5.6.1 Cement bound materials
There are a number of available materials that can be used successfully as a lightweight aggregate to formulate a low density core material. Table 5.4 gives a few materials and their corresponding air-dry densities. The following highlights the more commonly used materials.

Table 5.4 Air dry densities (loose) of lightweight aggregrates.

Aggregate type	Air dry density kg/m^3
Clinker	720–1040
Sintered pulverized-fuel ash	770–960
Foamed slag	560–960
Expanded clays and shales	320–960
Expanded slate	560–860
Pumice	480–880
Diatomite	450–800
Wood particles	320–480
Expanded perlite	80–320
Expanded vermiculite	60–160
Expanded polystyrene	<10

Perlite
Perlite is a volcanic rock found in a number of countries including Northern Ireland, Italy and America. When heated to 900–1100 °C it expands due to steam evolution and forms a cellular structure of 80–320 kg/m^3 bulk density. Table 5.5 gives a typical property value of

Table 5.5 Perlite concrete properties and mix proportions based on 100% yield (per m³).

Mix ratio (cement/perlite aggregate) by volume	Oven dry density kg/m³	Cement kg/m³	Perlite m³	Water m³	Air entraining agent m³	Wet density range kg/m³
1:4	576	376	1.0	0.23	0.0032	808 + 32
1:5	488	301	1.0	0.23	0.0032	728 ± 32
1:6	432	252	1.0	0.23	0.0032	648 + 32
1:7	352	188	1.0	0.23	0.0032	584 + 32

perlite concrete used as an insulating, lightweight core, together with the mix proportions for 1 m³.

Vermiculite
Vermiculite is a platy structure material found in America and Africa. It is similar to mica and, when heated to 650–1000 °C, expands up to 30 times its original volume. The bulk density of vermiculite is around 60–160 kg/m³. There is less compaction and higher yield when using the fine grades, and mixes made with these grades are usually more easily worked.

Table 5.6 includes a general comparison between vermiculite and perlite concrete.

Table 5.6 Typical properties of insulating lightweight concretes for use as core material.

Aggregate type	Mix by volume	Air dry density kg/m³	Min. 28 cube str. N/mm²	Thermal conductivity W/m °C
Vermiculite	1:8	400	0.7	0.094
	1:6	480	0.95	↓
	1:4	560	1.23	0.158
Perlite	1:7	400	1.40	0.098
	1:6	480	2.17	↓
	1:5	560	3.43	
	1:4	640	4.83	0.155
Polystyrene bead	—	200	0.29	0.081
	—	400	0.98	0.140
	—	600	2.16	0.204
	—	800	3.92	0.314
Graded wood particles	1:4	640	1.75	0.202
	1:3	880	4.90	↓
	1:2	1200	12.25	0.288

Polystyrene beads
Polystyrene beads are a German development sold under the trade name 'Styropor'. The type used is Styropor VP 358 which has a hydrophilic coating to aid the cement bond, unlike general polystyrene beads which have a hydrophobic coating. The raw beads have a short shelf life so it may be worth considering the installation of a bead expanding unit if large quantities are to be processed. Densities less than 400 kg/m³ are not recommended for core material in grc panels. A typical mix is given in Table 5.7.

Table 5.7 Typical mix using polystyrene beads.

1 m³ mix (density 450 kg/m³)
1 m³ of VP 358 beads
281 kg of oPc
98.5 kg of water
either 2.3 kg of Propiofan 325 D or Acronal 3003

Mixing procedure is now outlined:

1. Place the beads into the mixer and add the bonding aid dispersed into one third of the gauging water.
2. Sprinkle in one third of the cement and mix until the beads are fully coated.
3. Add the remainder of the cement and water.
4. Mix only to just ensure uniform dispersion without undue balling-up of the components.

Foamed concrete
One possibly suitable core material is foamed concrete. This is produced by adding a foaming agent, usually a hydrolyzed protein or resin soap, into a grout mix. During the mixing at high speed, air bubbles are stabilized and remain during placing, setting and hardening. Another process is to add a stable pre-foam to the grout while mixing in an ordinary paddle mixer. Hydrogen peroxide may also be added in this manner. Table 5.8 gives some typical properties of foamed concrete.

Table 5.8 Typical properties of foamed in situ concrete.

Dry density kg/m³	28-day cube strength N/mm²	Thermal conductivity W/m °C
288	0.5	0.08
495	1.3	0.10
640	1.9	0.11
800	2.9	0.17
960	4.1	0.23
1120	5.6	0.30
1280	7.7	0.42
1440	9.8	0.55

5.6.2 Board/sheet material

An effective lightweight core can be obtained by using sheet material. The obvious limitation is the lack of mouldability that rigid insulation boards can offer. They are, therefore, mainly used for flat composite insulation panels. The advantage given is that the rigid and resilient structure enables roller compaction to be easily carried out and the encapsulating grc layer sprayed over them.

Polystyrene and polyurethane have been used as a cast-in core and as a subsequently bonded core to cured grc sheets (Figure 5.3). When casting in these sheet materials, it must be made certain that air is not producing a void and subsequently preventing any bond between the core and the grc layers. Bonding adhesives are usually employed to ensure a consistent fix is obtained between core and the grc skins.

Figure 5.3 A grc composite panel formed by bonding cured grc to a polyurethane core.

Polystyrene

Polystyrene is obtainable in two forms:

- The fused together form of the expanded beads
- The extruded form of the above known as 'Styrofoam'.

The latter has greater rigidity and superior strength together with minimal moisture absorption and vapour permeance.

Large spans have to incorporate cross webs to strengthen sufficiently the composite panel. As mentioned before, bonding aids are required to ensure a uniform bond.

The normal stock profiles are sheets of rectangular section, but it may certainly prove cost effective to purchase blocks and cut them to production sizes using hot wire cutters. This can be done to produce the optimum usage and wastage may be bonded and recut.

The thermal conductivity of polystyrene is 0.033–0.04 W/m °C whereas Styrofoam is 0.029–0.033 W/m °C. The maximum operating temperature of polystyrene insulation is 80 °C and it should be remembered that solvents rapidly dissolve it. The stack must be covered if the material is to be stored in sunlight for more than one week. It should be noted that Styrofoam constitutes its own vapour barrier in its surface layers due to the nature of the material achieved after extrusion.

Polyurethane

Polyurethane has similar properties to polystyrene, but may be used as a structural core. The thermal insulation and cost are greater. The thermal conductivity varies between 0.017 and 0.023 W/m °C for a density of 34 kg/m^3. Flexible forms are also available. The working temperature range for polyurethanes is -150 °C $+$ 150 °C and dimensional stability is good over this range.

Perhaps the cost penalty of pre-formed board polyurethane makes it too expensive for the general use of a core material.

Resin bonded glassfibre

This form of core material is sold as cavity wall insulation for use during construction. It is available in 50 mm, 65 mm and 75 mm thicknesses of one size 1.2 m × 0.46 m which have a thermal conductivity of around 0.036 W/m °C. The limiting temperature under service is 230 °C. The material is not a water vapour barrier.

Roller compaction of the encasing grc skin will not be possible, but this may not cause concern in certain situations.

5.6.3 Spray-in foams

Although it may not be possible during production to spray a foam insulation (for example polyurethane) onto a grc skin while wet, in certain circumstances it may still be beneficial to use the bonding properties and rapid processing times of polyurethane. Grc skins that have been pre-made can be rapidly bonded to an insulating structural core to produce a component. This method may take the form of bonding two components of profiled or compound curvature, thereby producing one that would be uneconomical by conventional means. This use of polyurethane in an injection mode between two skins, is a recognized production process that has well-equipped machinery developed for its use. Due to the structural nature of polyurethane foams, this system of injection could be used to provide panels that have no cross web bracing and therefore no cold bridging.

5.7 MATERIALS WORTHY OF FURTHER INVESTIGATION

The following materials are noted as being worthy of investigation to determine precisely how cost effective they can be in grc production.

5.7.1 Suzorite mica

Suzorite mica is a thin, basal cleaved phlogopite flake

that is elastic, corrosion resistant, non-combustible and chemically inert. It is used as an inexpensive reinforcing agent in thermoplastics, thermoset resins, glass bonded mica, and cement boards, and as a filler in paints, plastics, glass, roofing materials and wall boards (see Figure 5.4). It is processed in a range of grades from a fine powder to a highly delaminated, high aspect ratio reinforcing flake as large as 6 mm. Supplies come from Quebec, Canada where some 30 million tons are available.

Figure 5.4 Illustrations of suzorite mica.

Suzorite mica has been used as a substitute for asbestos. Full scale Hatschek and Fourdrinier machines with minor alterations have processed material with a partial asbestos replacement. It is felt that its potential may lie in reducing the drying shrinkage of grc, increasing the flexural modulus and enabling the glass-fibre content to be lowered significantly. So far it is discovered that the dewatered sheet may be the best product to benefit by inclusion of suzorite mica.

5.7.2 Thermosetting cements

A Japanese thermosetting cement has recently become available in the UK market as Osaka T.S.A. This has the distinct advantage of setting in 30 minutes when heated at 80 °C. It can be demoulded and the products put into stock. The other advantage with this cement is that it can be used with 'E' glass or non-alkali resistant glass since it produces minimal attack on the fibre during hydration and curing. The elevated temperature has to be carried out in a sealed mould in order to stop rapid moisture loss that would occur in unsealed products heated to 80 °C. This prevents excessive shrinkage which could lead to cracking and crazing.

The ideal method for heating the moulds is to use a water bath or a steam tunnel. With such a rapid turn round expensive moulds can be used for many casts per day.

5.7.3 Regulated set cement

This cement is an American invention, which is marketed in Europe, but not to any great extent in the UK. Costs in early 1986 were £150 per tonne, against oPc and RHC at around £48 per tonne. The advantage given by the cement is that a controlled setting time can be achieved between 1 and 30 minutes. The cement consists essentially of a mixture of Portland cement and calcium fluro-aluminate ($C_{11}A_7.CaF_2$) with an appropriate retarder. Citric acid additions can be used to retard the set as well as chilled water.

The set achieved produces a low strength that may be useful to enable demoulding to take place. As normal hydration occurs with the Portland cement fraction of the cement, the strength is increased at the expected rate for oPc beyond the early strength achieved by the calcium fluro-aluminate. Naturally this cement has to be assessed as follows:

1. Can the processing equipment handle such a rapid setting cement without blockages or other problems?
2. Are the extra costs of equipping the plant to handle this cement justifiable?
3. Can an overall saving be made to the manufacturing costs of the product?

6

Mixers, mix designs and mixing procedures

The following Chapter gives background information on mixers, mix designs and mixing procedures. In the main, only what may be termed basic mix designs are detailed but, since these form the majority of mix designs used in production, they are worthy of individual attention. Firstly mixers used or suited for grc production are examined.

6.1 MIXERS

Mixers divide into three types — rotating pan, high shear and continuous.

6.1.1 Rotating pan mixers

This type of mixer, found in most precasting works, was initially used for mixing grc mortar before the benefits of high shear mixers were explored. Nowadays its role is in the mixing of lightweight aggregate mixes for use as core material in sandwich construction and also as a means of dispersing fibre in a pre-blended slurry (caused by high shear mixing) to produce premix composites.

The drum is driven at 16 rpm and the mixing blades, located off centre in the mid position along the radius of the drum, are rotated at 74 rpm. Hence one rotation of the drum ensures that contact is made between the mixing blades and the whole volume of the mix. The machine is available in two sizes, 100 litres and 50 litres capacity (Figure 6.1).

Figure 6.1 The Liner cum-flow rotating pan mixer.

6.1.2 High shear mixers

Present production of grc for spray-up and spray-up dewatered material realizes the benefits of high shear mixing which produces a well-dispersed uniform mix, free from lumps, in a rapid time (circa 1 minute).

Greaves high shear mixers

These mixers have evolved through collaboration between Pilkington Brothers and Joshua Greaves & Sons Ltd. The basic unit consists of an impeller fixed to a vertical shaft and connected to a vertically mounted 3 hp motor.

The Greaves 50 and 80 mixers are the most popular models. The Greaves 50 mixer has a cement bag holder and shoot. It is available trolley or wall mounted with alternative hydraulic or pneumatic lift. All rotating parts are guarded by an automatic electric isolating switch that operates when the mixer is raised. The largest mixer, the Greaves 150, operates by the same impeller and drive system. This machine is supplied with a tilting frame to assist discharge.

All the Greaves grc mixers employ a specially developed impeller (Figure 6.2). This is made from hardened steel in a disc shape, with radial cuts equally spaced around the circumference enabling alternate blades to be bent at right angles either up or down.

Figure 6.2 The Greaves mixing impeller.

The cement and sand are pulled down towards the impeller by the vortex action and as they traverse along the impeller blade are subjected to inter-particle shear. On leaving the blade they are exposed to high shear by the oncoming teeth. The particles are then injected at high velocity into the body of the slurry and the cycle repeats itself.

An adaption which enables the mixer to discharge directly into the mortar pump hopper is shown in Figure 6.3. Once mounted on a turntable bed, the mixer could then feed two pump hoppers located on each side of the mixer.

Tumac THB-100 and THB-50 mixers

A recently developed machine is the Tumac THB-100 high speed mortar mixer. This machine, shown in Figure 6.4, is a Swedish mixer of similar design to the Greaves. It has a 4 hp motor and a 100 litre capacity. The machine is provided with a stand that enables it to be discharged into the mortar pump. It also comes with wheels for transporting.

A smaller machine, the THB-50 is shown in Figure 6.5. As illustrated, the mixer can be attached to a tilting stand or supplied fixed to the mortar pump. It has a 50 litre capacity, 1.5 hp motor and a mixing cycle of 3 minutes.

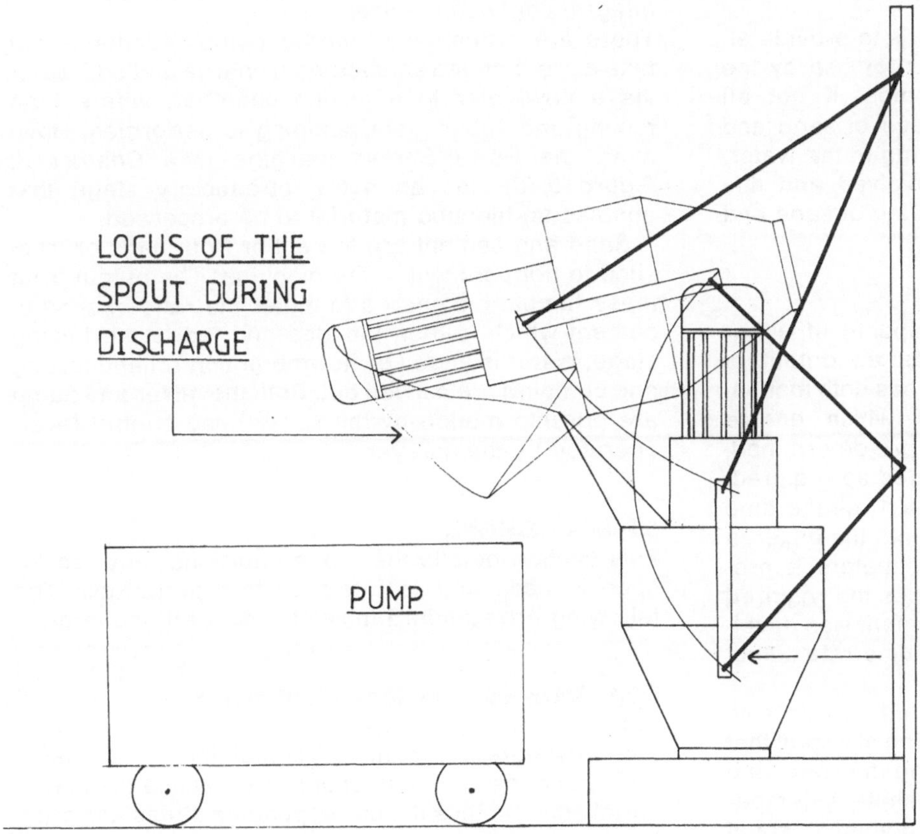

Figure 6.3 Mode of discharge of adapted Greaves 150 slurry mixer.

Figure 6.4 The Tumac THB-100 high speed mortar-mixer.

Figure 6.5 The Tumac THB-50 high speed mortar-mixer.

6.1.3 Continuous mixers

The principle of the continuous mixer is to provide an uninterrupted supply of mixed mortar for use by the spray pump or pumps. However, most if not all continuous mixers require the dry feed of sand and cement to be pre-blended before mixing in the water. A two-stage process is therefore involved and this should be borne in mind as pre-blending of sand and cement requires a large capital outlay.

Bancem continuous mixer

The Bancem continuous mixer is capable of either batch or continuous mixing and has an output of 5–35 kg/min. of mixed mortar. The unit has both manual and interlocked automatic controls which enable water/cement ratios to be accurately pre-determined, checked during operations and repeated as required. The various functions are related to a 1 minute time cycle. As pre-blended material is fed into the mixer, a pre-set sequence is activated which meters a pre-determined water addition and produces the required slurry formulation. Dry blending equipment was developed for this machine, working on the time/rate output principle.

The mixer consists of a tube, mounted horizontally, in which revolves an assembly consisting of a shaft that locates a number of discs spaced along the shaft, and fixed perpendicular to the axis of the shaft. A number of rods are welded to the disc at points of equal distance around the circumference and parallel to the centre shaft. When rotated, these rods act as the mixing blades and produce a shear action to the material input. In order to propel the mix along the mixing chamber, a helically wound rod, located around the rotating mixing assembly described above, is employed.

Greaves cement slurrying system

The Greaves cement slurrying system is illustrated in Figure 6.6. This system is based on the Greaves high shear mixer and is fed with sand and cement from screw feed silos. A water-gauging device is fitted. The process is a batch process.

Integral continuous mixer

There are a number of mortar pumps available that take a pre-blended sand/cement charge and add water via a flowmeter to a mixing chamber, where both mixing and subsequent pumping is performed. However, the Power-Sprays machine (see *Chapter 3, Figure 3.10*) has an extra introductory stage that enables un-blended material to be processed.

Sand and cement are fed either from bags or from silos to hoppers within the machine. The output from these hoppers is geared to meter the ratio of sand to cement which is then blended and fed to the mixing stage. Water is added by flowmeter control and mixing and pumping are carried out. Both the mixer and pump are put into motion by the spray head control lever, operated by the sprayer.

6.2 MIX DESIGNS

This section details the mix proportions required for specific uses and to attain certain properties. The following are standard mixes for general use in grc.

6.2.1 Spray-up (non-dewatered) matrix

Set out in the Table are three mixes for a spray-up (non-dewatered) matrix containing increasing amounts of sand. The intention is to reduce the drying shrinkage and thus the corresponding stress and crazing that can occur in under-sanded mixes. These mixes are offered therefore as possible alternatives (material specifications are given in *Chapter 5*).

	1	2	3
Cement	30 kg	25 kg	30 kg
Sand	10 kg	12 kg	30 kg
Water	⩽9.5 kg	8 kg	10.2 kg

NOTE
Or the equivalent proportions based on a 50 kg bag of cement

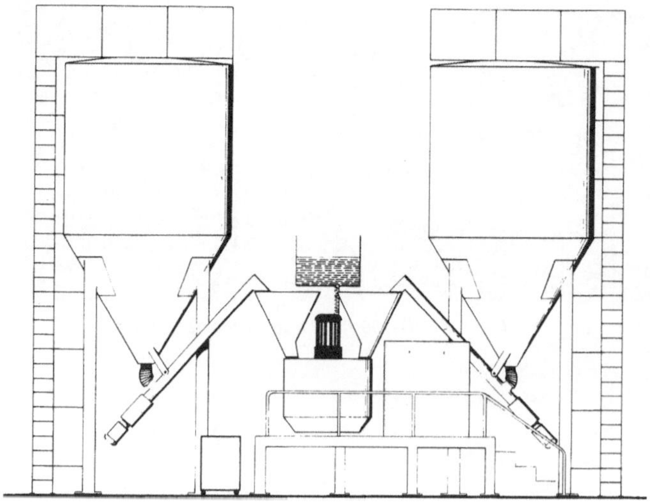

Figure 6.6 The Greaves cement slurrying system.

6.2.2 Spray-up (dewatered) matrix

Cement	30 kg
Sand	10 kg
Water	15 kg

6.2.3 Pre-mix grc

Cement	30 kg
Sand	10 kg
Water	10 kg
Cormix P.1.	150 g
or	
Cormix 75/24	30 g
Glassfibre 25 or 12 mm	up to 2.5 kg

6.2.4 Fire resistant mix for spray-up, non-dewatered grc

This mix is designed to improve stability and integrity up to 2 hours under fire test conditions but it does so at the expense of reducing density and general physical properties. For this reason, it should only be used internally. Further improvements to the insulation value enabling the requirements of BS 476: Part 8 to be met can be achieved by coating the surface with vermiculite mortar renders.

	Parts by weight
oPc or RHPC	100
Blended pfa	67
Water	47
Glassfibre	11.3
*Cormix 74/8 A.E.A.	0.4

*An organic sulphonate manufactured by J Crosfield.

6.2.5 Styropor lightweight core mix for sandwich construction

	1	2
Styropor beads VP 358	1 m³	1 m³
Cement	267	281
Sand	14	—
Water	98.5	98.5
Acronal DS 3003	2.3	2.3
or		
Propiofan 325 D		

6.2.6 Injection moulding premix

oPc/pfa Blend	60/40
Water/dry solids	0.45–0.48
Celacol	0.4%
Glassfibre (22 mm)	5%

6.2.7 Pressing mix, premix

oPc/pfa	50/50
Water/dry solids	⩾0.3
Celacol	0.025–0.25%
or	
Polyox	
Glassfibre	2.5–5.0%

6.2.8 Extrusion mix, premix

Ram extrusion through a die

oPc/pfa	50/50
Celacol	0.2%
Water/solids	0.45
Glassfibre (22 mm)	2.5%

6.2.9 Flat sheet extrusion

oPc/pfa	50/50
Water/solids	0.4
Celacol	0.8%
Glassfibre (22 mm)	2.5%

6.3 MIXING PROCEDURES

The following is a general guide to the sequence of loading and mixing of grc and associated mixes in grc production.

6.3.1 General mixing procedure for spray-up grc

1. Charge the mixer (a high shear mixer) with the gauging water and any liquid admixture specified.
2. Then progressively add the sand and cement. Alternatively add the sand first and then the cement, while mixing.
3. Any powder admixture that is specified should be blended with a portion of the sand inclusion.

6.3.2 General mixing procedure for premix grc

1. Mix the matrix component first in a high shear mixer as described above. This will produce a well-dispersed mortar that assists fibre dispersion.
2. Transfer the blended matrix material to a rotating pan mixer and gradually add the glassfibre. Only just sufficient mixing time is used to disperse the fibre and to prevent fibre 'balling' or 'build up' around the paddle blades.

6.3.3 Mixing procedure for styropor lightweight core material (rotating pan mixer)

1. Place the VP 358 beads into the mixer. Add the bonding agent with $\frac{1}{3}$ of the gauging water. Mix until the beads become fully coated.
2. Add about $\frac{1}{4}$ of the total sand/cement or cement requirement to the mixer and mix until the beads are fully coated.
3. Add the remaining cement and water.
4. Mix until full dispersion has occurred but before excessive 'balling up' has taken place.

7

Manufacturing methods and techniques

The following notes are intended to illustrate various methods requiring consideration for the manufacture of grc components. Information is given not only on techniques widely used but also on recent developments and new ideas. At times material contained in other chapters is of particular relevance and, when indicated, cross reference should prove useful.

7.1 PREMIX GRC

By definition, premix grc is produced by including the glassfibre at the mixing stage and dispersing the fibres in the matrix by the action of the mixer. This produces a 3D array of fibres within the wet composite and will consequently, for a given fibre content, provide a composite possessing a lower strength than spray-up grc.

Manufacturing techniques have been developed to improve the properties obtained by earlier methods, but premix should be considered only suitable for components with relatively low performance requirements.

7.1.1 Production mixing method

Initially the premix process relied on dispersing the matrix and fibres simultaneously in a mixer of the forced pan type. Recent developments, however, favour separate mixing of the matrix followed by the introduction of the fibres and sufficient mixing to cause wetting-out but not filamentization.

The matrix is blended by using high shear mixing to prepare the matrix and overcome agglomeration of the cement particles. This dispersion may be assisted by using deflocculants, usual admixtures being carboxy methyl cellulose and polyethalene oxide.

The process is as follows:

1. The gauging water is placed in the vessel of a high shear mixer, together with the admixture (if used).
2. The cement is added and mixed for 2 minutes.
3. The sand is then added and mixed for a further 1 minute. It is important to note that an increase in mixing time causes the mixing temperature to rise and can lead to an acceleration of the initial set of the mix.
4. The blended matrix is then transferred to a paddle mixer and the fibres sprinkled in quickly but carefully to avoid 'clumping'. The mixing time is kept to a minimum to disperse the fibres and cause sufficient 'wet-out' of the composite. This is crucial because fibre damage is greater with premix grc. Excessive over dispersion mixing can cause strand breakdown into individual filaments reducing the MOR value, 'E' value and impact properties. This strand filamentization, even when only slightly occurring, leads to noticeable reductions in the ability of the mix to flow into confined spaces. Mix flow is also a function of the fibre length but when fibres shorter than 12.5 mm are used a rapid fall off in properties occurs.

7.1.2 Moulding

The mould should not have sharp arrises or corners. It must be of a suitable dimension to enable the mix to flow into the mould retaining distribution and 3D orientation of the fibres to prevent realignment. If realignment occurs due to inadequacies in these respects or because the mix is too dry having a restricted flow, the result will be a loss in strength of the composite. Similarly too wet a mix can cause segregation during compaction.

7.1.3 Compaction

The most economical compaction method is by vibration. The mould is clamped to a vibrating table with, ideally, a vibration of 9000 c.p.s. Filling points should be kept to a minimum to avoid interbonding weakness and the prevention of a uniform flow of the composite into the mould. Experiment is necessary to optimize the number of filling points.

Vibrating the mould and trowelling may be beneficial as well as employing a weighed contoured plate on top of the composite. Vacuum dewatering may also assist compaction. Cost considerations will, however, of necessity decide the method that is eventually employed to suit the particular shape of the product.

7.2 HAND SPRAY-UP GRC

Hand spray-up of grc provides a most versatile and cost effective method of producing grc components. A whole range of profiles can be made including flat sheet components, profiled components fabricated in either male or female moulds and composite components including hollow sections or lightweight core materials.

By this means a good quality material is produced. It ensures efficient usage of fibre and minimizes fibre damage during the manufacturing process. A basic investment in production equipment is offset by easy adaption to produce a whole range of components. Naturally, manning levels will be high when related to square metreage output, but versatility in the type of products that can be continually produced may be of more importance to the manufacturer.

When specific products have been identified it may well be found that a specialized automated process handling greater outputs at reduced manning levels is then preferable to the hand spray-up process.

7.2.1 Basic procedures

Before considering production procedures in detail the reader is advised to be familiar with the contents of *Chapter 2: Requirements for production* and *Chapter 3: Production equipment*. With a basic knowledge of the production layout it is possible to proceed through the stages of a simple hand spray-up process.

7.2.2 Slurry preparation

When considering the method suitable for producing

the cement and sand slurry or matrix, two important factors should be kept in mind:

1. The minimum amount of water must be employed in the mixing process in order to achieve satisfactory workability. Excess water will result in lower strengths, greater shrinkage, possible shrinkage crazing and greater permeability of the composite, causing increased wetting/drying and dimensional instability.

2. The components of the matrix (sand and cement) are variable materials that may well require changes in the mix proportions, or the use of admixtures, in order to produce a suitably workable mix, with the minimum water content.

The mixing procedures and suggested basic mixes are detailed in *Chapter 6*.

7.2.3 Setting-up the equipment

The spray pump, mixer and roving box should be positioned so that the operator has the minimum of discomfort in performing the spray-up process. This means that the slurry hose is either held by a boom or so arranged as to avoid continuous movement when traversing the mould with the spray head. The air hoses should be taped to the slurry hose again to minimize discomfort. They should also be colour coded to allow quick reassembly after unavoidable maintenance during manufacture.

The fibre roving must be supported in such a manner as to facilitate 'take-up' by the chopper gun and prevent contamination by overspray or dust en route. Similarly the roving box requires storage in a suitable area away from damp, overspray and impact damage.

Equipment calibration

In order that the percentage fibre output of the equipment can be set, it is necessary to determine the individual outputs from the mortar spray head and the chopper gun. This is simply done by weighing separately the outputs from the two spray heads, over a set period of time.

The required fibre content for the composite will be known from the specification and therefore the output levels can be set by reference to *Chapter 11: Figure 11.4*. This graph breaks down a range of fibre contents into the amounts of fibre and cement required to achieve these percentage fibre inclusions.

It should be noted here that the fibre content is defined as:

$$\%\text{ fibre inclusion} = \frac{\text{fibre content}}{\text{fibre content} + \text{mortar content}}$$

This method of specifying the fibre content leads to the series of non-parallel lines representing the fibre contents in *Figure 11.4*.

The procedure is therefore to set the mortar pump output at a level that is within the capacity of the machine, for example 13 kg/min. Then one should read vertically up the graph from the 13 kg/min. position on the x axis, to the intersection with the specified fibre content line. From the intersection of these two lines, the corresponding fibre output is read on the y axis, to determine the fibre output required to achieve the specified fibre content. For 13 kg of mortar output at 5% fibre content, this would require therefore 680 grams of fibres.

Experience gained over a number of production runs will enable the operators to determine just how much above the level of fibre output requirement the fibre output must in fact be set to ensure the specified requirement is met. This is necessary because of:

1. Variations in the drag on the roving during manufacture

2. The non-consistency of take-up of the chopper gun and such like during spraying.

A good starting level is an excess of 0.3–0.5% of the specified fibre content, and this requires setting the fibre content in the given example to between 5.3% and 5.5% or, from *Figure 11.4*, at 750 g/min. of fibre output.

In conclusion, when the output of fibre has been decided, it must be determined that the chopper gun can in fact operate at that particular level and give a consistent output (refer to *Chapter 3, Section 3.2.1* and *Figure 3.15*). This only becomes a problem when operators set the mortar output up to a high level in order to quicken the spraying time. It is then found that fibre contents obtained from wash-out tests reveal a variation and lower fibre content than is specified.

Spray-gun laydown

Before components can be sprayed-up with the confidence that a uniform 'wet-out' and 'laydown' pattern is being achieved, a test should be made on a piece of card or polythene. Reference to *Chapter 3: Figure 3.12* shows the basic details of a hand spray head. With the mortar spray head vertical and maintained at right angles to the surface at a distance of 300–400 mm, two passes across are made, one with the chopper gun leading and the second to one side of the first, with the chopper gun trailing. The ideal laydown is achieved when uniform mixing of the fibre and mortar occurs before depositing on the surface, thus ensuring no 'dry fibre' is visible. If dry fibre is being laid down then the inclusion angle may require adjustment and the spray pattern again examined as before. The mortar spray cone may also need alteration by either relocating the air nozzle (as is possible on the earlier mortar spray heads) or by adjusting the air pressure to the mortar spray head.

It may be noticed that the fibre is being bounced out of the mortar spray cone and this can be remedied by either lowering the air pressure to the mortar spray head, or increasing the air pressure to the chopper air mover. The inclusion angle α should not require changing once an optimum setting has been found, and trimming air pressure settings should enable the spray pattern to be finalized. Misalignment of the mortar spray cone and the chopped fibre spray can sometimes occur in the plane of the illustration *Figure 3.12*. This should be checked by viewing the laydown process in action. The modern spray heads are not likely to cause this unless nozzles have worn irregularly and require replacement. The older type of spray heads, as shown in *Figure 3.19*, which use a connecting plate to locate the mortar spray head and the chopper gun unit, have been known to suffer from this problem. Similar problems have occurred with the traversing head unit shown in *Figure 3.14*.

7.2.4 Component manufacture

Component manufacture involves the following stages which are now examined — pre-spraying checks, test board manufacture, spraying, finishing, demoulding and making good.

Pre-spraying checks
Naturally it is hoped that the majority of making good and mould checking would take place prior to moving the moulds into the area adjacent to spraying. These checks can be carried out either at the demoulding and making good stage or during mould cleaning and oiling (as shown in *Figure 2.1*), and would be the responsibility of the production Q/C laboratory staff. However, a basic assessment of the mould should be made by the spray team in advance of spraying. Recycled moulds in particular require special attention. These tests should ensure:

1. The mould has been assembled correctly (good mould design can make certain that this becomes foolproof).

2. The mould oil or release agent has been applied satisfactorily and excessive ponding has not occurred in recesses for example.

3. Overspray, dust and the like have not contaminated the mould surface.

Test board manufacture
Early on in the preparation of the first product at the start of a shift a test board should be sprayed-up and roller compacted by the same labour force using exactly the same methods. Samples are taken from the test board to determine the glass content and water/solids ratio and hence the production output. The remainder of the test board is then cured and tested in accordance with the details in *Chapter 11*, for strength and density properties.

Spraying procedure
It is customary first to apply a 'mist coat' of pure sand and cement slurry to the mould. This is intended to pick up the surface details of the mould and prevent the subsequent fibre addition from appearing at the surface and reducing the moulding of any surface detail. The mist coat is usually a single pass of the mortar spray head. Often, in order to reduce crazing of the surface of a grc component, a polymer addition of an acrylic or S.B.R. latex is added to the mist coat formulation. In this case it may be desirable to spray the mist coat with a hopper-fed texture surface gun.

Once the surface of the mould has been coated with the mist spray, the main spray-up of the grc must follow directly in order to achieve an indistinguishable bond between the mist spray and the grc backing.

The grc is sprayed by making straight lines across the mould face, with the mortar nozzle perpendicular to the mould face, and at about 200 mm from the surface of the mould. Often it is found beneficial to spray into the corners and return details of the mould first, and compact these, then following on with the bulk of the spray work.

Layers of 3 mm thickness are laid down and then the material is compacted with rollers, to remove any air voids. Further layers of 3 mm are laid down and compacted until the design thickness has been achieved. Thickness checks are made, as described in *Chapter 11*, and low thickness areas are sprayed over to produce a minimum equal to the design thickness.

Corners tend to become bridged with dry fibres during the spraying process and should be well compacted with a special tool. *Chapter 4* gives details of these tools and also the techniques used for component spray-up. These techniques should be understood and implemented. The following points are worthy of particular note:

- Never use overspray to assist in the filling of moulds
- Remove all dry glass build-ups from corners etc.
- Do not transfer sprayed material from one location to another in the mould as bonding will not be formed between the layers.

Finishing
Wherever possible, overspray can be turned into the mould to achieve a good edge detail. Surplus material can then be cut off when 'green' or allowed to set before removal with a sharp knife. Following this the back face can be trowelled flat and the overspray around the mould perimeter removed before setting. Sometimes it is required that the back face receives a rollered finish. This would be performed while the grc is still green.

The moulding tools are best stored and returned to a bucket of water adjacent to the mould during spray-up to keep them clean and serviceable. Dirty tools tend to pick up the composite during compaction and trowelling.

When the product is finished, the back face is covered in polythene and transported to the overnight curing area, which is maintained at 20 °C or thereabouts.

Demoulding
Products are usually demoulded between 16 and 24 h after spraying. Naturally care must be taken during this operation since the stress induced while demoulding can be critical and lead to product breakages and failure. Practical mould design combined with the correct release agent reduce the risk of damage. Grc products are, however, in the main thin planar products. This contrasts with concrete which is a bulk product and consequently the demoulding team should exercise the necessary care.

Making good
The most important aspect about making good is that it should be done as soon after demoulding as possible, and certainly before the products are put in for final cure. It must be emphasized that some defects cannot satisfactorily be made good and these are usually caused by excessive voidage in the matrix. Places of particular concern are around fixing points and in areas that will be highly stressed. The usual amount of voidage that is found on demoulding, however, is easily made good with the normal matrix proportions of sand, cement and water and is preferably boosted with a polymeric admixture to improve bonding.

7.2.5 Plant cleaning and maintenance
The quality of the output obtained from a grc production facility is directly dependent on the manufacturer's attitude towards plant cleaning and maintenance. Sufficient time for example must be set aside in the schedule to allow for both to be carried out.

At the end of each shift the production equipment must be washed down. All the components that come into contact with the grout, such as the pump hopper and filter, pipes, stator assembly, spray gun and nozzle, should be included. At the same time checks can be carried out for wear. After this the equipment needs reassembling correctly ready for the next shift.

The two most common problems that are incurred during spraying are:

1. Blockage of the mortar spray system due to flakes or lumps of grout entering the delivery pipe and nozzle.
2. Inconsistency of output from the chopper gun. This is usually the result of maladjustment or worn components. The cutter blades are best replaced at the end of each shift (or sometimes sooner depending on output rates). The backing roller should also be checked for wear and the pressure that it exerts on the cutting blades requires setting whenever the cutting blades are replaced.

7.2.6 Plant lay-outs and output rates

Lay-out of any plant depends on the space available and the production process in mind. Hand spray grc can be carried out in four basic ways:

- Static mould
- Rotating mould
- Production line moulding
- Station moulding.

Static mould

In this process the mould is brought to the spray area and deposited in front of the operator, who then moves around the mould spraying-up the component. Rolling is carried out during the spraying but final finishing is performed after the product has been removed from the spray area.

Rotating mould

With this system, the sprayer benefits from the placing of the mould on to a revolving turntable that is either mechanically indexed round or can be turned by hand. In this way the sprayer need not move and therefore is not struggling with heavy delivery hoses and such like.

Consideration should be given to the setting up of two turntables, one either side of the sprayer. This would enable spraying to be carried out on one while roller compaction or finishing takes place on the other.

Production line moulding

This technique was developed for cladding panel manufacture. A linear carriage or roller table carries the moulds in a conveying system to each stage of the process. Typical stages are:

1. Spray main outside shape into mould
2. Move along to filling location
3. Fill with core material
4. Move to second spraying location (or cycle back to 1.)
5. Spray final back coat and trowel
6. Move away for curing.

Station moulding

This method is similar to *production line moulding* in that each particular process is carried out at a set location in the factory. However, in production line moulding, the sequence is laid out in a line along a conveyor system, whereas in station moulding the sequences are located at various points around the production area, and the mould is transported from one to the next by either carrying, fork lift, overhead crane or hand barrow. This may mean that once set-up the production facility is cheaper and more flexible.

7.3 AUTOMATED SPRAY-UP GRC

When a particular product has been decided upon after market research it becomes obvious that production costs can be reduced by automated production facilities.

Automated production of grc is a vague term and can vary between merely replacing the function of the sprayer by using an automatic traverse unit to setting-up a complete facility. Various techniques have been investigated and proven for a range of products from flat sheet to cladding panels and concrete pipes. The equipment used in such processes is discussed in *Chapter 3*.

It is worth mentioning here that when a rapid and proven system is developed and successfully employed the benefits of using a rapid setting cement should be considered. This rapid set can make early handling and demoulding a relatively simple process and cheaper than endeavouring to struggle on with oPc formulations. One such cement is known as Regulated Set Cement, which is available in Germany and the USA. This cement can be formulated so that its own internal chemistry can produce a set at any time from 5 minutes to 45 minutes. Another way of inducing the set is to initiate the reactions by infra-red heaters. This causes the set to occur at a particular stage in the process which then becomes predictable and consistent. The cement is expensive but the rapid turn round of moulds and product output rates have led a few manufacturers to tool-up production to utilize this material.

7.4 EXTRUSION

This process has been developed in order to produce products from premix material. The equipment has been discussed in *Chapter 3*. A range of products have been developed using the extrusion technique, but production has not as yet developed sufficiently to make the process cost effective. The end result has, however, shown little benefit, if any, over that of spray-up production.

7.5 PRESSING

Pressing grc is best considered as a stage in the production process of either spray-up or premix. It should be considered as a method of compaction rather than a total technique. It is possibly better suited to the processing of premix material and more likely to be of benefit in the production of cast products rather than sheet, 2D components.

7.6 VACUUM FORMING

As with pressing, vacuum forming is best considered as a stage in the production process, more likely to be suited to spray-up work.

Vacuum forming can be utilized to produce products from male or female moulds. Normally female moulds would be used in downwards vacuuming (as carried out for example on vacuum tables). Upwards vacuuming would usually employ male moulds, that would be drawn up into a backing structure.

Besides compacting the grc it also removes excess water and produces a good quality dense product with superior properties.

7.7 WINDING

This process consists of winding the reinforcement round a rotating mould, which determines the internal dimensions of the product and the internal surface profile. The reinforcement may be in a variety of forms; rovings, matts or fabric (Figure 7.1A). It is drawn in a taut condition through a bath of cement paste, which is continuously stirred to prevent segregation by pumping it through a mortar pump, and returning it under pressure to the bath.

This system can be adapted further by winding the reinforcement round a cylindrical mould or roll. At the same time, chopped strands are sprayed onto the revolving mould as well as cement paste. The composite is compacted by means of compressing rollers (Figure 7.1B).

The advantages of this system are:

1. It enables different types of product to be made from sheets taken from the drum in the green state (flat sheets, troughs, pipes and shells).

2. Production can be easily mechanized to produce a consistent quality of grc.

3. It can employ continuous rovings that can be uniformly wound onto the mould, thus ensuring firstly a uniform dispersion of the reinforcement throughout the composite, and secondly the most efficient use of the fibre.

4. The fibre content that is obtained by this method, due to its uniformity throughout the product, yields a composite that has consistent strength values throughout.

7.8 PERIODIC DELAMINATION

This method is suitable for making thin-walled structures, for example shells, irrigation channels and embossed units, as well as sheeting. It operates on the following principles:

1. A glassfibre mat is placed between two planes and immersed in a bath of cement paste. A certain amount of paste therefore penetrates into the mat.

2. The planes enclosing the mat are then moved further apart. The cement paste wets both surfaces of the mat and partially penetrates into it.

3. The planes enclosing the mat are now compressed onto the mat. Its external layers are impregnated with cement paste which will adhere to these planes so that, when the latter again moves apart, the mat will segregate along those layers of fibre which the paste has not yet reached.

By repeating this procedure many times, the glass mat becomes fully impregnated with cement paste, producing grc composite.

This principle has been used to develop two methods for manufacturing grc products, the first with application for units such as shells and irrigation channels, the second for producing flat sheets.

Manufacturing method for shells and irrigation channels

This application is detailed in Figure 7.2. It consists of a matched metal die moulding unit and a unit for supplying the cement paste. The bottom of the die is joined to a longitudinal container from which the

Figure 7.1 Diagrams showing grc fabrication by winding.

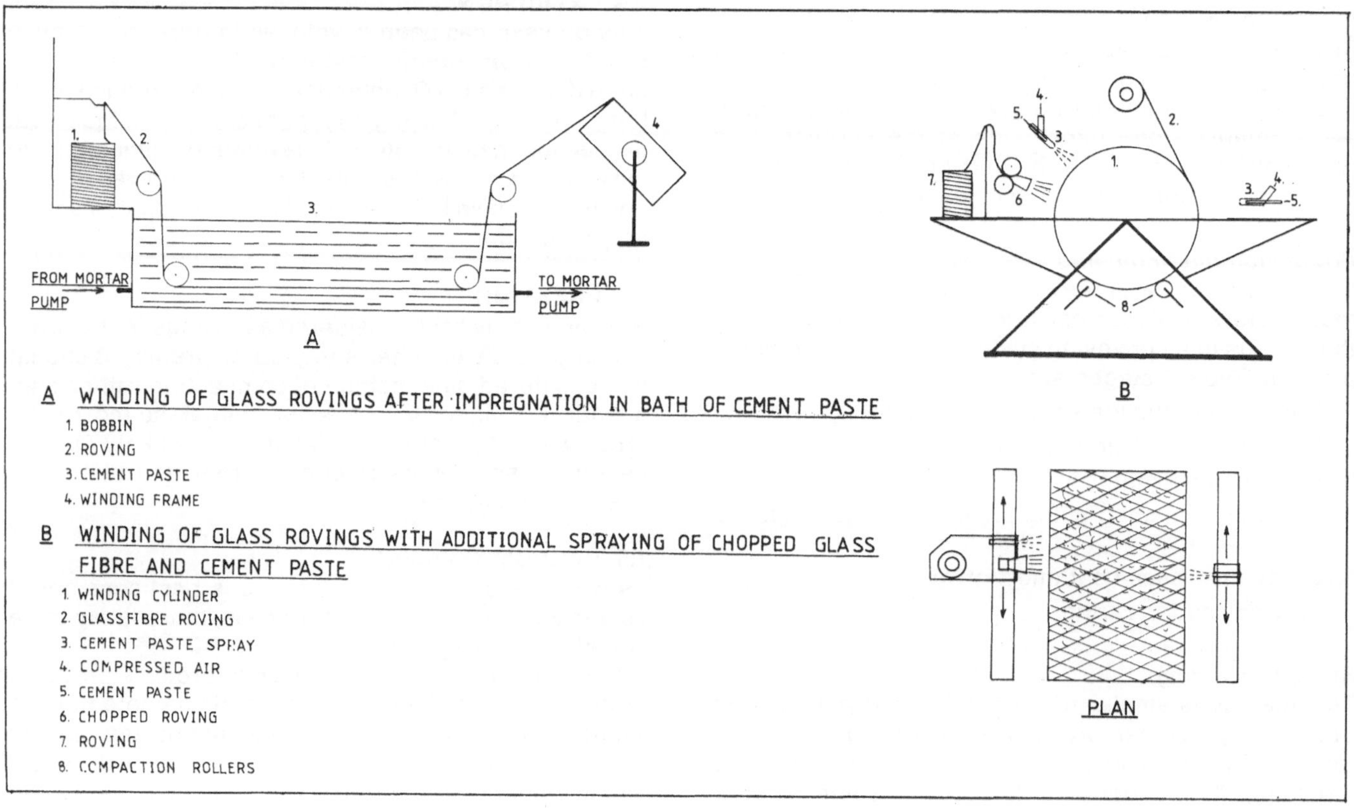

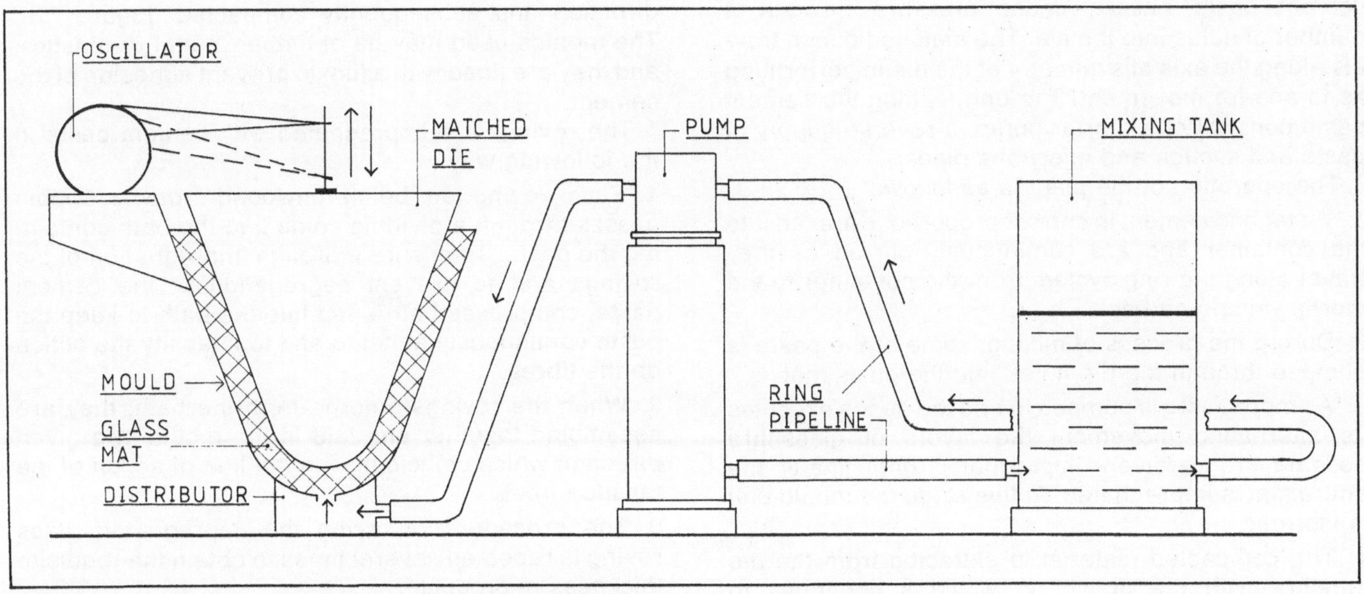

Figure 7.2 An illustration of the periodic delamination method of impregnating a glass mat to fabricate curved grc products.

Figure 7.3 Diagrams showing the manufacture of grc window frames by continuous moulding of glassfibre rovings impregnated with cement paste.

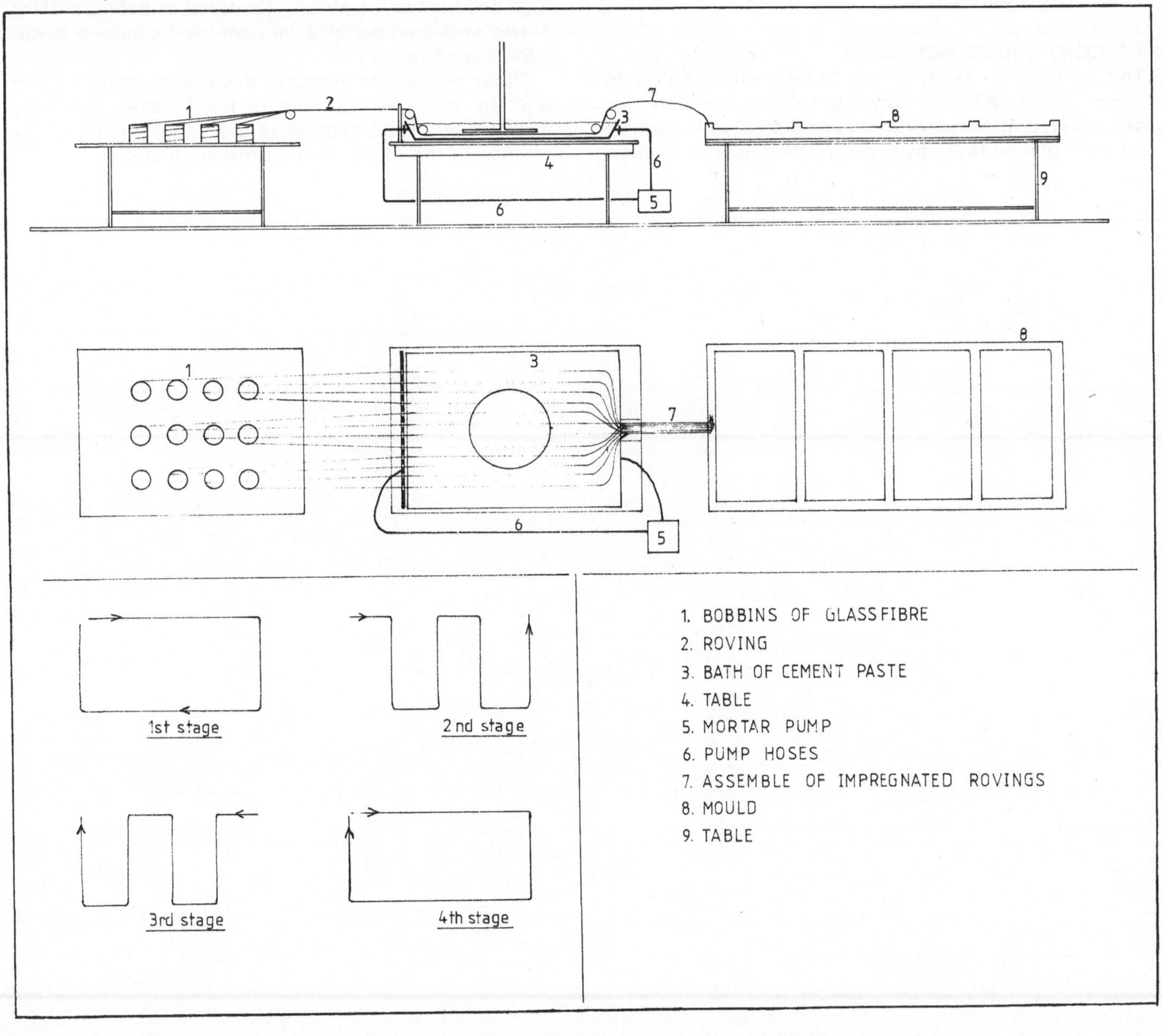

cement paste passes under pressure through a number of holes into the die. The matched punch travels along the axis of symmetry of the die in performing its to-and-fro movement. The unit feeding the cement paste consists of a mortar pump, a reverse supply of paste and suction and injections pipes.

The operation of the plant is as follows:

1. Water and cement in correct proportions are fed into the container and are continuously stirred as they travel along the ring system from the container to the mortar pump and back.
2. During the process of mixing, some of the paste is pumped through the die holes into the glass mat.
3. As more paste is pumped in and the punch executes its alternating movement the layers of glassfibre become progressively impregnated until finally the entire mat is impregnated. At this stage the mould unit is vibrated.
4. The compacted material is extracted from the die together with the punch on which it continues to harden.

The advantage of this system is its high output rate combined with its simplicity of design. Its main drawback is the need to have a large number of punches since each is immobilized while the product adhering to it is being cured.

7.9 CONTINUOUS MOULDING

This method is ideally suited to thin-walled products which are small and variable in cross-section. An assembly of glass rovings impregnated with cement paste is placed layer by layer in the mould in a given direction and subsequently compacted (Figure 7.3). The moulds used may be of timber, metal or plastics, and they are lined with a film to prevent adhesion of the cement.

The rovings are impregnated with cement paste in the following way:

1. Each roving on being unwound from a bobbin passes through a dividing comb into the bath containing the paste. To ensure thorough impregnation of the rovings and to prevent segregation of the cement paste, compressed air is fed into the bath to keep the paste continuously agitated and to intensify the action on the fibres.
2. When the rovings emerge from the bath, they are assembled together and laid in the mould in a given direction which coincides with the line of action of the tensile forces.
3. The procedure of laying the impregnated glass roving is repeated several times to obtain the requisite thickness of product.
4. Essential fittings are placed in the mould during the moulding process.
5. When placed in position in the mould, the rovings are compacted by hand.

The strengths obtained by this method are uniformly high because of the strictly observed orientation of the fibres and the use of a cement paste having a low water/cement ratio.

Other noteworthy features are the low labour costs and the good appearance of the finished products. Conversely, the drawback is the hand placing of the rovings in the mould and their compaction.

8

Curing

Carefully controlled curing is vital to the production of good quality, serviceable grc which will attain strength levels required in the specification and reduce early shrinkage to a minimum. Grc is, by no means, exceptional in this respect. Basic concrete technology illustrates the significance of curing.

8.1 GENERAL EXPLANATION

The term 'curing' requires clarification. It applies to procedures used to promote the hydration of cement, which can only take place in water-filled capillaries, and consists of the control of temperature and moisture movement from and into the grc. Any loss of water by evaporation must be prevented, and internal loss by self-desiccation must be replaced from outside. The hydrating cement should, therefore, be kept saturated or as nearly saturated as possible, until the originally water-filled space in the fresh cement paste has been filled to the required level by the products of hydration. Figure 8.1 shows the order of influence of moisture on strength.

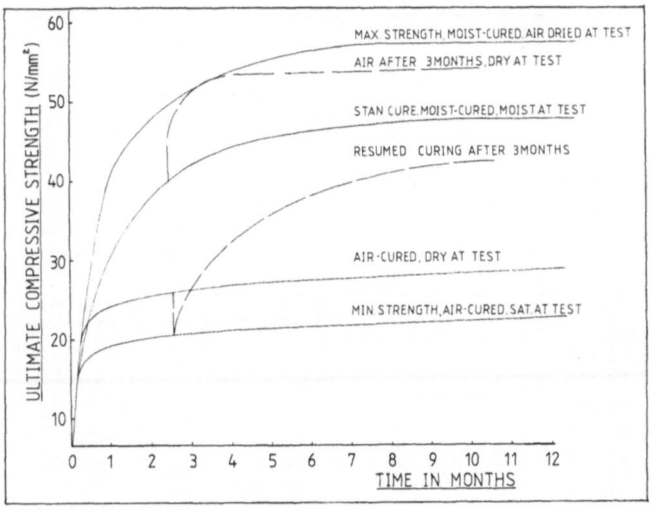

Figure 8.1 Variations in concrete strength due entirely to curing conditions and state at testing.

The effects of the curing environment must also be seen in relation to the condition at the time of testing, i.e. that water cured material tested dry is stronger than water cured material tested in the saturated state. The full range of apparent strength levels, therefore, lies between dry cured material tested saturated and water cured material tested dry (refer to Figure 8.1).

8.2 SUCCESSFUL CURING

To ensure that ingress of water into the grc takes place the following environments have been successfully employed:

1. A curing chamber or building with a consistently maintained environment of 95–100% RH at 15–20 °C
2. Shrink wrapping with a surplus of water within the wrapping while maintaining a temperature of 15–20 °C
3. Total immersion into a water tank again at 15–20 °C.

It should be realized that hydration can only proceed if the water present in the cement paste is at least twice that of the water already combined by hydration. Self desiccation is of importance in cement with water/cement ratios below 0.5. For higher water/cement ratios, the rate of hydration of a sealed specimen equals that of a saturated specimen. Grc has water/cement ratios in the range 0.28–0.35.

When the correct conditions for curing are provided, hydration occurs in a predictable manner as can be seen in Figure 8.2. The reactions start immediately after the water is added to the cement but significant effects begin to occur some 2–5 hours after the initial mixing. The eventual degree of hydration is however dependent on:

Time and water/Cement ratio

The longer the time and the higher the water/cement ratio, the better the degree of hydration which is reached. The S shape curves shown in Figure 8.2 are typical of reactions in dilute solutions. This confirms that the hydration is a true chemical reaction.

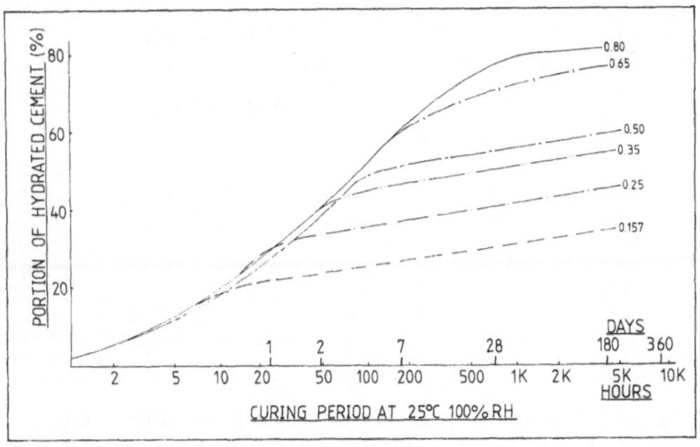

Figure 8.2 Progress of hydration in cement pastes versus time.

8.3 MATURITY

Reference should be made to the maturity concept. For particular mix proportions cured under a maintained humidity level, the strength gain is a function of:

Time × Temperature

and this summation is called maturity. The temperature is ascertained from an origin found experimentally to be between −12 and −10 °C (take −11 °) and is therefore expressed in °C-hours.

Strength plotted against the logarithm of maturity gives a straight line for cement pastes. This relationship holds true for the limit of proportionality, bend-over point and modulus values, but not necessarily for the modulus of rupture, ultimate tensile strength and impact values since these are dependent on fibre/matrix interaction.

8.4 STRENGTH DEVELOPMENT UNDER 'NORMAL' CURING 20 °C/100% RH

When grc is cured at 20 °C/100% RH, which would relate to curing in a fog-room or by covering with polythene sheeting immediately after spraying, the strength development profiles shown in Figure 8.3 would be expected.

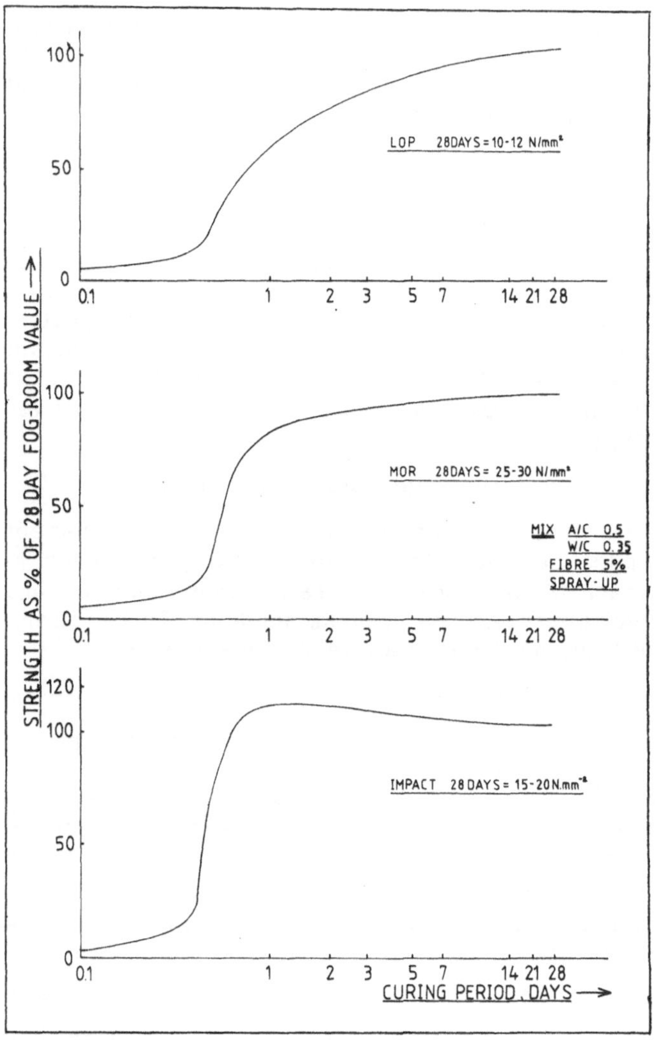

Figure 8.3 Strength development profiles for grc cured in fog-room at 20 °C/100% RH.

The LOP curve follows the pattern of the hydration curve in Figure 8.2 because it is a matrix dependent property. Therefore the maturity concept would hold true for LOP development. At 7 days, 90% of the 28-day strength value is attained within the composite. The MOR development is more rapid and 95% of the 28-day value is reached after 7 days. The impact strength development is quite different from the preceding values in that, under these curing conditions, the peak value is attained after 1 day followed by a drop off in value to a plateau at 28 days.

The previous strength development profiles are affected by the fineness of cement and degree of compaction. This is especially true during the first 24 hours. The drop off in the rate of strength development after the first 24 hours is a function of the w/c ratio used in grc.

It can be shown that at the water/cement ratios employed, hydration products give rise to a deposition of gel within the capillaries of the cement paste causing segmentation of the capillary structure. The segmented capillaries are then only interconnected by gel pores which slows down the rate of water ingress and possibly the hydration rate. This system is again a function of the water/cement ratio of <0.35. At normal concrete water/cement ratios of say 0.5, this segmentation process would take 6 months to establish.

8.5 STRENGTH DEVELOPMENT UNDER ACCELERATED CURING

The maturity concept mentioned earlier showed that strength development was a function of time and temperature. Therefore it would seem possible to increase the rate of strength gain by increasing the curing temperature. The validity of this can be seen in Figure 8.4. There are four factors that appear to affect the measured strengths in the material. These are:

1. Time of commencement of the heat-treatment
2. Duration of the heat-treatment
3. Temperature of the heat-treatment
4. Rate of temperature rise.

Figure 8.4 Strength development profiles of heat-treated grc compared with fog-room cured material.

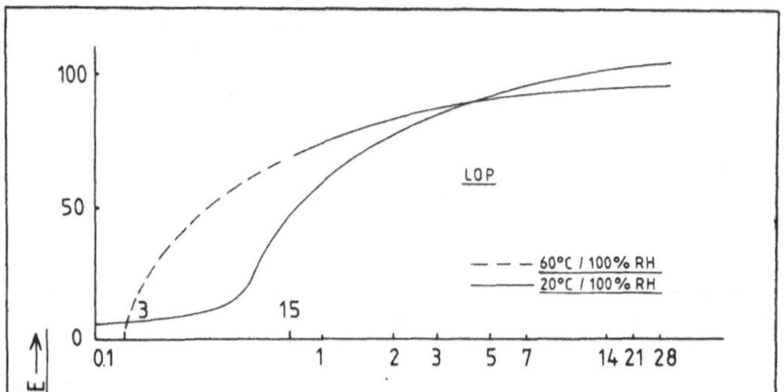

These factors have been identified in the accelerated curing of concrete, as can be seen by the procedures adopted by the ready-mixed concrete industry to carry out accelerated curing. However, in grc the factors that relate to the fibre/matrix interaction, i.e. MOR, UTS and impact values would yield a relationship that differs from a pure maturity consideration. The elevated temperature chosen was 60 °C since this relates to a temperature that would be easily attained by steam curing and may well be chosen for grc production. Figure 8.4 shows the normal curing profiles superimposed onto the response given by steam curing at 60 °C. The delay period chosen was 3 hours after mixing and spraying, and the treatment was applied for periods ranging from 3 to 21 hours. The rate of temperature rise was approximately 20 °C per hour.

The graphs show that, for the particular cement used (RHPC), the optimum period of treatment was 12 hours for both hydration and strength. After 1–5 days, however, depending on the property and probably the cement, the advantage disappears. After 28 days the heat-treated grc is, in comparison with 20 °C cured grc, 10% to 15% weaker in LOP and modulus, 20% in MOR and up to 35% in impact. Curing cement paste at elevated temperatures brings about changes in the microstructure of the hydrating cement that may cause strength variations.

8.6 CONSIDERATIONS BEFORE EMPLOYING ACCELERATED CURING SCHEMES

The following considerations arise from experiments that have been carried out on grc as well as practice that has been shown acceptable for concrete.

1. The grc product should be covered immediately after spraying to prevent moisture loss.
2. Curing should take place at 100% RH.
3. Heat treatment should take place within 3 to 5 hours of mixing the matrix.
4. The maximum temperature within the product should not exceed 50 °C.
5. The heating rate should not exceed 10 to 15 °C per hour.
6. The duration of the heat treatment should not exceed 12 hours.
7. The product should be kept at 100% RH for a further 6 days following the heat treatment to allow optimum strengths to develop.
8. The precise cooling rate after heat treatment will depend on the product section, but clearly units containing insulating core materials will heat up quicker than solid thick sections and also slow down quicker, hence distortion may occur.

It may well be worth considering heat-treatment curing only for simple single skin components due to the risk of distortion. Also, careful consideration must be made of the fact that the curing cycle will probably require manning during night hours and may prove awkward and expensive to operate.

8.7 EFFECTS OF CURING UNDER ADVERSE CONDITIONS

Figure 8.5 illustrates the effect of removing grc from a curing environment of 20 °C/100% RH at ages between 1 and 28 days after initiation of the cure. This particular case considers removal to a dry, warm environment such as that found in the Middle East. The conditions used for test were 35 °C/40–50% RH. Naturally windy conditions would assist in quickly reducing the free water content of the grc and bring about a faster reduction in strength.

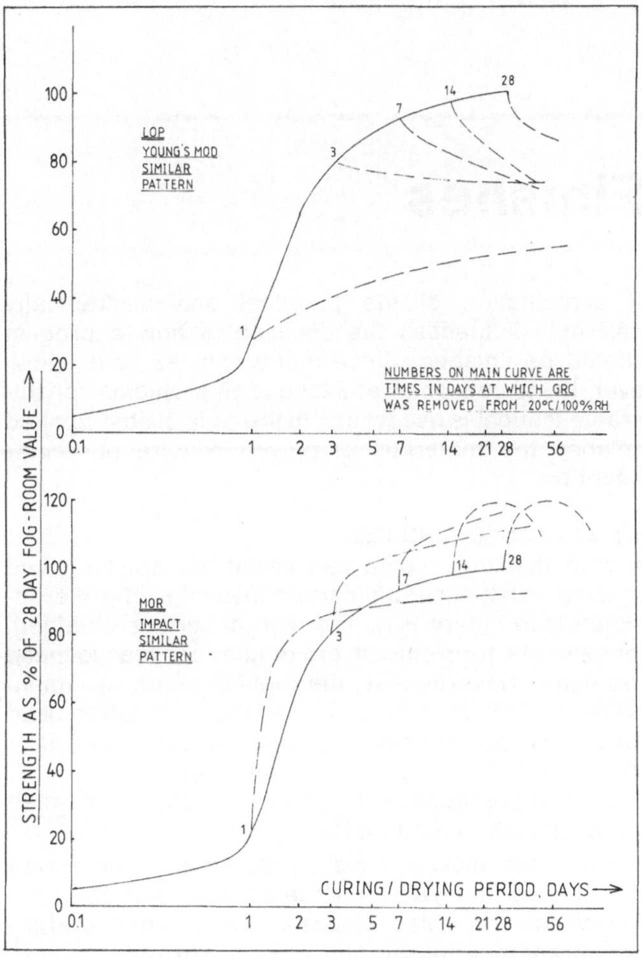

Figure 8.5 Effect of drying at 35 °C/40–50% RH on strengths of oPc cured at 20 °C/100% RH drying for up to 28 days after removal from fog-room.

The main effect noted is that the LOP values are reduced as related to the 28-day normal cured material (20 °C/100% RH) whereas the MOR values are increased. It is thought that the main effect of the LOP values is caused by the initiation of surface shrinkage microcracks. The 1-day-old material however was able to accommodate the environment change at this early age probably due to the continued presence of a certain amount of free water within the matrix. Material cured for 7 days at 20 °C/100% RH and then subjected to a drying environment of 35 °C/40–50% RH showed a reduction in the 28-day LOP value of 65–80% obtained from 28-day fog-room cured grc.

The effect on MOR values as mentioned is the reverse. Removal to a drying environment increases the MOR value. This is probably due to a shrinkage effect on the matrix which causes a more positive grip to establish itself on the fibres; leading to greater load transference between matrix and fibres.

In conclusion it is apparent, therefore, that the properties of the material should be carefully monitored following the establishment of a particular curing regime, to discover the full range of effects.

Finishes

A combination of the practical and aesthetically desirable influences the decision of how a product should be finished. Since many finishes have, however, led to both short and long term problems considerable thought is necessary. In the case of post-applied finishes the materials employed require particular attention.

Figure 9.2 Ex-mould finish taken from a rubber moulding of a natural stone wall.

9.1 EX-MOULD FINISHES

A smooth mould finish can assist the cracking and crazing which occurs in certain instances of grc production (see Figure 9.1). The type of surface obtained, for example, from smooth grp moulds can lead to these problems. Nevertheless, the moulded surface cannot be held totally responsible when the formulation used for the bulk of the composite was lacking in sand and/or the water content was excessively high.

Certain precautions are, however, advisable when using smooth faced moulds:

1. Minimum thickness mist coats should be employed possibly including a polymer latex such as an acrylic in the formulation. This provides a 'flexible' surface which is tolerant to a degree of shrinkage stress.
2. The mould should, if possible, be roughened.
3. The demoulded surface should be sealed with a proprietary sealer.

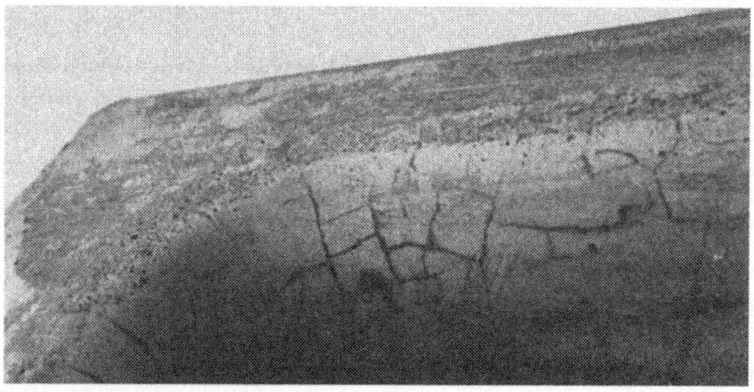

Figure 9.1 Crazed grc finish, assisted by the smoothness of the grp mould to promote this type of crazing.

With these precautions in mind, the potential of spray-up grc to take on the texture of a detailed mould is of considerable benefit to a manufacturer. Figure 9.2 illustrates excellent results obtained with a rubber mould taken from a natural stone wall.

Many textures are possible, as would be customary with precast concrete. If the lightness of grc is to be exploited, however, the relief of the finish should be restricted. On occasions when thickness and lightness are not critical the potential of grc to copy in detail a high relief profile can be used to full advantage.

9.2 LIGHT ETCH RETARDED FINISHES

A pleasant stone textured finish can be obtained by employing face-down retarders. These are used together with a mould release agent and are suitable for timber, wood, steel and plastic moulds. The result is a transformation of the glossy, craze-prone finish obtained from grp moulds to a sandy texture finish that removes the cement rich surface from the grc and exposes the sand in the matrix. It may, however, be necessary to increase the thickness of the mist coat in order to prevent exposure of the glassfibres on the etched surface.

The mould release agent is first painted on the mould and left to dry. Then the retarder is applied and allowed to dry out. Filling of the mould can follow at any time up to 5 days after the application of the etch retarder. A range of etch retarders is available giving a variety of depths but the lightest is probably the most satisfactory.

The application of the two pack system requires some consistency and care but the finished result, especially on white grc, is most impressive and represents Bath and Portland stone texture.

9.3 PIGMENTED FINISHES

The range of pigmented finishes that has been developed for ordinary concrete should be avoided in general for grc. Grc formulations contain such high percentages of cement that the rate and extent of lime bloom can be considerable. Consequently the dark pigmentation of grc will be inconsistent and spoilt by random and extensive patches of white lime bloom deposits. It is advised, therefore, that the most consistent finish is white grc and if pigmentation is required it should be restricted to the light pale shades.

9.4 EXPOSED AGGREGATE FINISHES

The normal techniques of exposed aggregate work are applicable to grc. The following points, however, should be borne in mind:

1. Small aggregate sizes are desirable with a maximum size of 10–12 mm. This keeps the weight down.
2. If retarded face-down work is considered, care should be exercised ensuring that the retarder has not been absorbed by the grc resulting in reduced strength when demoulding takes place.

9.5 STAIN FINISHES

A successful method for producing a pigmented surface on grc is to use a stain after demoulding. The proprietary stain available has an acid included in the formulation to clean off the top layers of cement and mould oil from the surface of the product. The solution is simply brushed on and left. Two applications are usually applied followed by a coating of silicone wax or liquid sealer to prevent lime bloom. This finish, if sealed carefully after application, has proved quite successful and practical to use. The buff stain is probably ferrous sulphate solution and the green is copper sulphate, both assisted by an acid in the solution.

9.6 SEALERS

A range of proprietary sealers is available with varying influences on the surface of the grc. The purpose of a sealer should be to penetrate the surface and reduce significantly the take-up of atmospheric dirt and pollution. This may mean that the surface appearance is changed by the sealer but a range of effects can occur from a deep gloss to a barely noticeable sheen.

9.7 EPOXY SPRAY APPLIED FINISHES

Epoxy resins are perhaps the most successful materials to bond onto concrete. One disadvantage with epoxies is their lack of resistance to ultraviolet light degradation. The application of a surface of polyurethane over the epoxy can, however, achieve a tolerable gloss finish. It should be emphasized that the system cannot be stated as totally reliable until proved by exhaustive development work. The author has, however, carried out a certain amount of investigation which so far confirms that the system has potential. Figure 9.3 shows a section of a window frame coated with an epoxy that is not resistant to ultraviolet light. The bond is totally acceptable and indeed this particular formulation is used to line food vats.

The equipment required for this application operates by heating the resin and hardener and metering the required ratio separately along heated pipes to the spray head where a turbine mixer and nozzle assembly sprays the mixed hot resin. This system has the advantage that the curing resin, when contacting the grc, cools rapidly which reverts the fluid resin back to its thixotropic ambient state. By this means vertical surfaces can be sprayed without run marks occurring.

Figure 9.3 Window frame section coated with a spray applied epoxy finish.

9.8 EPOXY GEL-COAT FINISHES

The bonding of an epoxy resin finish by gel-moulding has many distinct advantages.

1. The finish is governed by the quality and surface of the mould.
2. The texture and shine of the finish obtained is constant.
3. Relatively complicated profiles can be produced.
4. Semi-transparent motifs and writing can be bonded into the surface.

The process involves coating the mould with a poly vinyl alcohol release agent as used in grp moulding work. A special two-part epoxy resin is then applied by spraying or painting to form a thick uniform coating. Once 'tacky' a second thin coating is added again by either means and this acts as a bonding layer. The grc is then sprayed immediately onto the two-coat system and the process completed as standard. When the grc is sufficiently hard enough to resist demoulding the unit is removed.

Timing is critical in this process. The degree of curing permitted for the first coating of epoxy is most important since an insufficient allowance would encourage the grc to push through the two layers and appear in patches on the product surface. Over-cure of the first layer would prevent binding to the bond coating between the surface and the grc. Delamination can then occur.

There are, therefore, a number of stages which have to be followed systematically to achieve results but this coating offers great potential in producing colourful, glossy finishes to grc. It has had considerable use for street furniture and a selection of these items are shown in Figure 9.4. Other applications have included

Figure 9.4 Range of street furniture coated by gel-moulding the epoxy coating to the product during manufacture.

shower cubicle panels and, after 4 years of use, a trial panel has shown no signs of deterioration.

9.9 PVC SURFACE COATINGS

One manufacturer has developed a technique of surfacing grc sandwich panels employing a PVC sheeting with apparently good results.

9.10 ALIPHATIC POLYURETHANE TWO-PACK FINISHES

Initially, considerable success was obtained with this coating. General acceptance and use of this material however led to instances of unexpected bubbling on the surface at times from 1 day to as much as a few months. Roller application was thought to counteract this risk but in general the finish has now fallen out of favour.

The cause of the breakdowns in bond were thought to be affected by the surface moisture content of the grc together with the mould oil remaining in the grc surface after demoulding. As a coating for cladding with the risk of interstitial condensation it is obviously not to be recommended. The bubbles formed in the film coating were found to contain water and as they spread and increased in size the film eventually peeled off completely. This breakdown may be a result of osmosis at the coating/grc interface.

Figure 9.5 shows a range of litter-bins that were coated satisfactorily with the material. Figure 9.6 shows a finial for the Regent's Park Mosque that was eventually covered successfully with the finish.

Figure 9.5 Street furniture coated with an aliphatic polyurethane two pack finish.

Figure 9.6 One of the finials of the Central London Mosque, coated with an aliphatic polyurethane two pack finish.

9.11 ACRYLIC POLYMER BASED TEXTURED FINISHES

These applied finishes are totally reliable and satisfactory as coatings to grc. A variety of colours and textures can be obtained and many manufacturers offer products in this range. The texture obtained is easily achieved and can be varied to give a series of possibilities, as shown in Figure 9.7. This finish can also conceal a limited amount of surface defects.

Figure 9.7 Range of textured finishes obtained with acrylic polymer based coatings.

9.12 IMITATION BRICK RENDERS

A convincing brick effect can be obtained by a two-part render. The grc is first coated with a bonding polymer latex. This is then covered with a pigmented render of 5 mm thickness of the chosen pointing colour for the brickwork. Above this is applied a 10 mm rendering of the selected brickwork colour which is left to harden. The last render contains an integral retarder which allows a special cutter to be used providing the required brickwork shapes 24 hours or so after application. The cutter passes through the brickwork render and just into the underlying pointing render. Figure 9.8 shows the effect obtained. The brick texture is produced by a gentle stippling of the top render, with a paint brush, immediately after applying the render.

Figure 9.8 Two coat rendered brickwork effect.

9.13 POLYMER IMPREGNATION COATINGS

A durable glossy coating can be achieved by the application of a pre-polymer syrup of methyl methacrylate. The coating requires full curing to obtain optimum results and complements the stain colouring described in 9.5.

10

Design

The Pilkington Brothers *Design Guide* provided much useful background information and is a good starting point for the designer employing grc. Sections highlight cladding panels such as fixings, joints and sealing, and surface finishes. There is, however, the need for a more rational approach to design. The following proposed design method is the work of Fordyce and Wodehouse (Building Design Partnership) and takes account of current developments in Codes of Practice relating to structural engineering design. Much, however, remains to be learnt and unfortunately this can only be achieved with experience gained in use.

10.1 DESIGN ASPECTS

Every product has to be designed whether it is a litter-bin or a shell roof. In the first instance someone has to decide that the product is required, what its function is to be, and how it should perform in carrying out that function. The material to be used and its size, shape, thickness, surface finish, colour and so on have then to be determined along with any other details of the product such as jointing and fixing methods. A designer is concerned with all these aspects to a greater or lesser degree depending on the product and its application. The approach to designing in grc is no different from any other material and must involve all these basic steps.

Thus design can be seen to lie between research and manufacture, and the designer must translate the theories of the one into practical design methods which take into account the deficiencies of the other.

An important subject at present is the question of the structural use of grc. The licensees of the Pilkington 'Incorporation Licence' Agreement that process Cem-FIL AR glassfibre, operate on the understanding that primary load-bearing and fully structural applications will be avoided until sufficient knowledge is available on the long-term performance of grc.

10.2 ANALYSIS OF GRC SECTIONS

Having considered the general factors affecting a particular application, the designer has to examine the actual behaviour of the component under applied loads in its environment. The Pilkington Brothers *Design Guide* uses a working stress approach for feasibility calculations coupled with prototype testing. It is essentially an extension of the trial and error method and again should be satisfactory, provided the designer has studied the material and realizes the limitations of the method.

There are two basic problems with this approach. Firstly, working stress levels appear to be arbitrarily set and no attempt is made to correlate these with particular material strengths, nor to make allowance for the effects of the environment or the particular application. Secondly, in prototype testing of products with time-dependent properties there is the problem of determining the long-term behaviour from tests carried out at an early age. Thus a more rational approach to design is required especially as the move is made towards structural uses. It then becomes necessary to design with economy and to be able to predict reliably long-term behaviour.

From studying the basic materials, the constituents and properties, the methods of manufacture and the influence these have on the properties and having considered how the material is used in practice, a basis for design can be established.

Although grc is not a single material and covers a range of materials with properties dependent on the particular formulation used, most information is available on certain standard mixes which form the bases of those currently employed. However, long-term data are only available on material with a neat cement matrix made by the machine spray and dewatering process. The main properties and factors which need to be considered are:

1. Strength properties of grc
2. Shrinkage, moisture and thermal movements
3. Manufacturer's limitations
4. Material properties and control tests
5. Behaviour of full size components
6. Associated materials.

The design method has to incorporate these factors into a logical framework and should also be such that, as more knowledge and confidence are gained about the materials, refinements and adjustments can be made.

10.2.1 Strength properties

For most applications the designer is particularly concerned with bending behaviour, as bending tension will generally be the critical stress. The limit of proportionality in bending (LOP) is adopted as the basic for the design method, rather than the modulus of rupture (MOR) for the following reasons:

- There is insufficient knowledge of the stress/strain behaviour of grc up to failure, and the MOR is therefore not meaningful as a measure of the actual stress in the material at failure.
- Below the LOP, simple bending theory is applicable.
- The designer is mostly concerned with the avoidance of cracking in the matrix.
- The LOP is less variable and its behaviour in different environments is more predictable than the MOR. It should be noted, however, that there is a variation between the LOP results as measured by different operators and instruments, but this is a function of the test method rather than the material behaviour.
- At stress levels below the LOP, fatigue and stress rupture will not be critical.

Although the design is based on the LOP, the

pseudo-ductile behaviour of grc above the LOP up to the MOR provides the ultimate factor of safety against failure. The loss of this pseudo-ductility and the fall in the MOR in wet or natural weathering conditions thus reduce the ultimate factor of safety from its initial value. This is illustrated by the MOR/LOP ratio for neat cement material which typically is about 2.5 to 3.0 at 28 days. This ratio is maintained for dry conditions but falls to about 1.2 to 1.3 for wet conditions (Figure 10.1). The effect of sand additions on the durability has given rise to little published data; however there appears to be some long-term benefit.

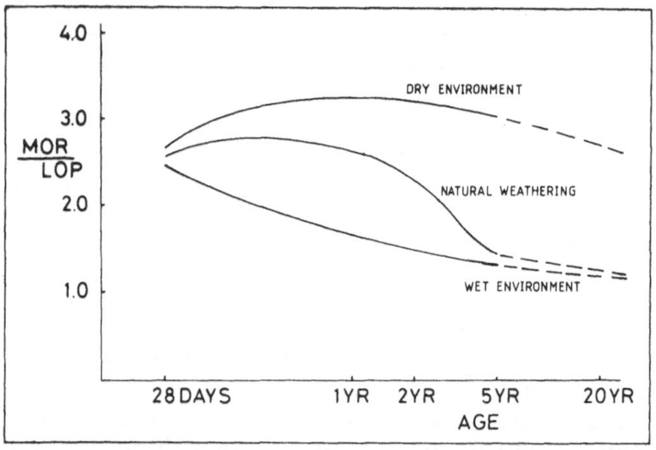

Figure 10.1 MOR:LOP ratio (average values, neat cement matrix).

Thus, there are two factors to be considered. Firstly, in dry conditions the LOP value falls from the initial 28-day value to a minimum of about 60–70% of that value. Secondly, for wet conditions and natural weathering, although the LOP remains at or above the initial value, a reduction factor must be applied to maintain an adequate reserve of ultimate strength.

Although bending tension has been taken as the significant property for design, there are situations where the direct tensile strength becomes critical. This is the case for box sections or sandwich construction. Thus the bend-over point (BOP) becomes the basis for design with the ultimate tensile strength (UTS) providing the reserve of ultimate strength. By considering the LOP:BOP and MOR:UTS ratios, reduction factors may be applied for cases where the strain gradient tends towards a more uniform tensile stress across the section.

10.2.2 Shrinkage, moisture and thermal movements

When movements are restrained, tensile stresses will be induced in grc. The magnitude depends on the degree of restraint and the stress/strain characteristics of the material at the particular age under consideration. When the ultimate drying shrinkage of a standard grc mix, typically about 0.15%, is compared with a tensile cracking strain under immediate loading of about 0.05% it is clearly very important to consider the factors which would restrain the shrinkage movement.

The restraint may be external or internal from variations within the material, variations in drying rates, incorporation of other materials and such like. The effect of the restraint has to be considered with the value of Young's modulus for the material at the particular age together with the reduction which will occur through creep in tension. Consequently the effect of restraint on these movements is very much reduced.

Although little information is available at present for these properties of grc, assessments can be made from studies of neat cement mortars and concretes. It is apparent that these stresses are not secondary stresses and they may indeed be of a similar magnitude to those resulting from applied loads and could lead to cracking in the matrix.

10.2.3 Manufacturers' limitations

The design method has to recognize the difference between material produced by the manufacturer and that produced in the laboratory. It has been observed that where the manufacturer's control is good the material produced compares well with the laboratory material but poor control leads to lower and more variable results. In addition, there may also be differences between the material in the actual product and the test board used for quality recording. These may arise from the spraying process or from the curing method. The differences may therefore be put into two categories:

- Normal variations which arise in any manufacturing process
- Other variations which are particular to grc and mainly due to fibre orientation and the dispersion and curing effect.

Account must also be taken of possible variations in skin thickness which may be undetected in the product. With thin sheet materials, small variations can cause significant stress increases and it has been observed that large variations are possible.

10.2.4 Material property and control test

The combination of fibre content, LOP and MOR testing is required for proper control of grc production to give a check on the adequacy and strength of the material being produced—matrix, fibre content and composite. Normal statistical methods may then be used to establish characteristic values to be employed in design and to check compliance during manufacture. These are detailed in the next Chapter.

10.2.5 Behaviour of full size components

Account must be taken of the difference in behaviour of the full size components made for a particular application compared with the standard test coupon in bending. This is essentially a question of assessing strain gradients across the section under consideration and taking account of size and shape effects.

10.2.6 Associated materials

These may include infills, surface coatings, fixings and jointing materials. The interaction of grc with infills is of particular importance. Naturally where they are used as a structural element in the design, the interfacial bond with the grc and the stress/strain characteristics of the infill must be known.

10.3 DESIGN METHOD

The aim of any design is to achieve an acceptable probability that the unit being designed will not become unfit for use during its specified life. It should not fail. Difficulty arises, however, in defining failure. It may be defined more precisely as that state in a structure or product when it ceases to fulfil the function for which it was designed and is deemed to have reached a limit state. There are two categories of limit state:

- Ultimate—which corresponds to the maximum load condition causing collapse
- Serviceability—which relates to criteria governing normal use; for example, cracking, unacceptable deformations or other damage.

At this stage in the development of grc there is insufficient knowledge and understanding of its behaviour to adopt the ultimate limit state approach. This is particularly so with regard to the long-term behaviour of cracked grc in bending, tension and shear, and the interaction of these stresses with those resulting from shrinkage, moisture and thermal movements. However, an approach can be adopted based on the serviceability limit state modified to take account of the known ultimate behaviour.

Thus this design method adopts the limit of proportionality (LOP) in bending as the basis for design. The factors to be taken into account are grouped under these coefficients or partial factors of safety:

γ_m—factors which affect the strength of the product

γ_f—factors which affect the load or load effects on the product

γ_c—a coefficient used to modify the design values in appropriate cases to take account of the modes of collapse and the consequences of failure.

The particular values selected for the material properties are related to 'standard' grc mixes, however the design method has general application. Standard material is defined as containing cement to BS 12—fine silica sand (zone 4) at 20–25% by weight of total mix, water (to give a water:cement ratio of 0.28–0.33) and glassfibres (nominal 5% by weight, Cem–FIL fibres 34–38 mm long).

To satisfy the design, the load effect in the material has to be maintained less than the load resistance of the material. The limit state is reached when they are equal, i.e. $F_d \leqslant f_d$ where F_d is the design load or load effect and f_d is the design strength.

Design load and design strength are decided from the characteristic values. These are determined statistically to take account of variations in the load and strength during the life of the product.

The characteristic load value (F_k) is taken from those set out in CP3, *Chapter 5, Loading*. This value has to be increased by the partial factor γ_f to take account of the variability of load or load effects. Thus $F_d = \gamma_f F_k$.

The characteristic strength value (f_k) is based on results obtained from four-point bending tests to determine the LOP in bending (*Figure 11.17*). It is therefore necessary to reduce this value by the partial factor γ_m to allow for variations in the strength of the material in the product. Thus $f_d = \dfrac{f_k}{\gamma_m}$.

The relationship of design and characteristic values for load and strength is illustrated in Figure 10.2.

In addition stresses arising from restrained shrinkage and thermal movement must be included in compilation of the design load stress.

10.3.1 Partial factors

The load increase coefficient γ_f is taken as the product of three coefficients which account for the variability of

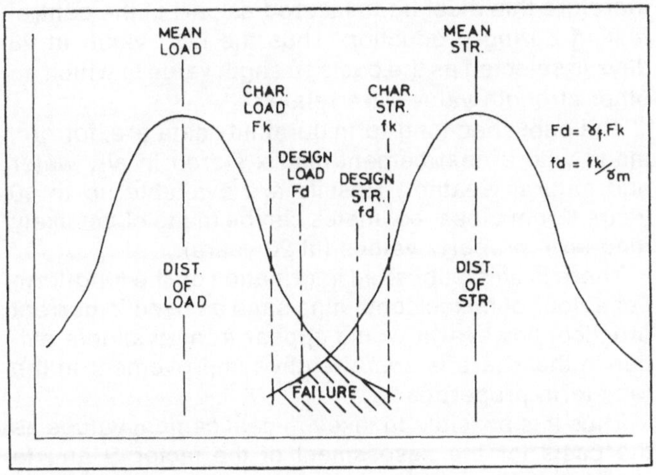

Figure 10.2 Relationship of design and characteristic values for load and strength.

loads and load effects, i.e.,

$$\gamma_f = \gamma_{f1} \times E_1 \times E_2$$

γ_{f1} takes account of the variability of load over and above those catered for in the characteristic load. The appropriate factors are those from CP110 and range from 0.9 to 1.6 depending on the load and condition being considered.

E_1 takes account of the sensitivity to thickness variations in grc, which would cause a stress increase over and above that allowed for in the compilation of γ_f in CP110. Values range from 1.0 to 1.2 depending on manufacturing method and complexity of the product.

E_2 takes account of the difference in behaviour of the full size component compared with the test coupon. It is related to the effects of stress gradients across the section and values will range from 1.0 to 1.5 depending on the overall depth and configuration of the section.

Strength reduction coefficient γ_m is taken as the product of two coefficients which accounts for the variations in the strength of the material in the product, i.e.,

$$\gamma_m = \gamma_{m1} \times D$$

γ_{m1} takes account of the reduction in strength of materials in the product as compared with the control test specimen and local variations in strength within the product. This factor incorporates the difference between the test coupon strength and that of an equivalent coupon taken from the product, being dependent on the complexity or otherwise of the product covering:

- Variations in strength between mixes
- Variations in curing between the product and the coupon
- Local variations caused by randomness of fibre orientation and dispersion.

There is insufficient data available to assess this factor for grc. It should be less than that adopted for concrete but, until it is demonstrated that manufacturers can achieve consistent standards of production, γ_m is taken as the value for concrete of 1.5. For automated processes where better control is possible, a lower factor could be taken (1.3 or 1.4). *D* takes account of the strength reduction of grc with time in different environments.

The design is based on the LOP in bending and

therefore this must be measured as part of the control testing during production. Thus the LOP value at 28 days is selected as the basic strength value to which all other strength values are related.

The published long-term durability data are, for grc, made with a neat cement matrix stored in air, water and natural weather. Results are available up to 10 years. From these, estimates can be made of the likely long-term property values (at 20 years).

There is little published information on the long-term behaviour of mixes containing sand as used in current practice; however it would appear from available evidence that there is some relative improvement in the long-term properties.

Thus it is possible to take the neat cement values as the basis for the assessment of the factor D and to modify this for sanded mixes. In addition, for simplicity, water storage and natural weathering are considered together, the strengths being similar at 5 years with natural weathering having the lowest predicted long-term values.

Air stored material

A simplified presentation of the published data for air stored material is given in Figure 10.3(a), all values being related to the LOP at 28 days.

The lowest value of LOP is obtained at 2 to 3 years, followed by a small upward trend to 5 years and is assumed to level off thereafter. The MOR shows a steady but relatively slow fall from the 28-day value. If the minimum LOP is selected as the basis, then this will give a value for D of 1.6 $\left(\text{i.e. } \frac{1}{0.63}\right)$. However, this results in a large strength reserve with a final MOR:LOP ratio of 3.0 $\left(\text{i.e. } \frac{0.9}{0.63}\right)$. It is more reasonable therefore to take the long-term value for D of 1.4 $\left(\text{i.e. } \frac{1}{0.7}\right)$ and to accept a temporary reduced factor of safety against matrix cracking at 2 to 3 years. This value for D gives a final MOR:LOP ratio of 2.7 $\left(\text{i.e. } \frac{1.9}{0.7}\right)$ which is the same as the 28-day value and gives an adequate reserve of ultimate strength. If it is assumed that sanded mixes will show improved performance, D may be reduced to 1.25 $\left(\text{i.e. } \frac{1}{0.8}\right)$.

Figure 10.3a Evaluation of strength reduction coefficient D—grc in air (all values related to LOP at 28 days. Average values, neat cement matrix).

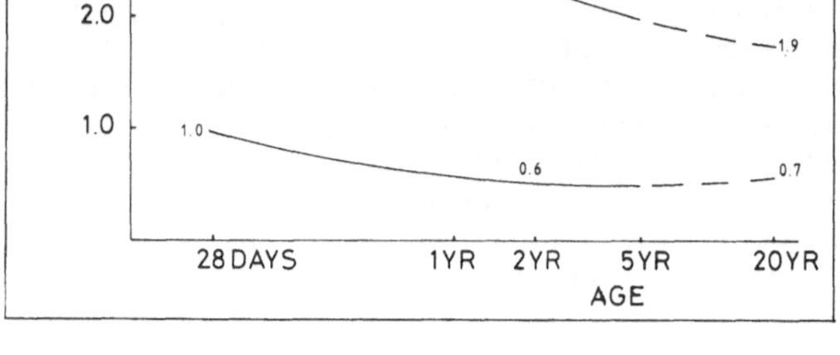

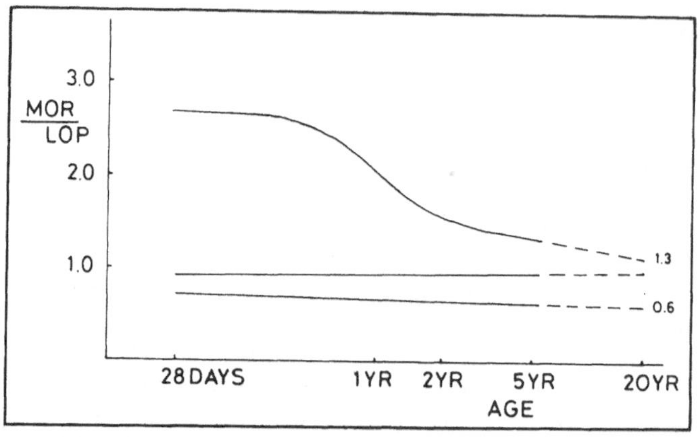

Figure 10.3b Evaluation of strength reduction coefficient D—grc in natural weathering for UK (all values related to LOP at 28 days. Average values, neat cement matrix).

Natural weather

Figure 10.3(b) gives the simplified presentation of the published data for natural weathering, again all values are related to the LOP at 28 days. In this case the LOP shows a slight increase from 28 days to 5 years, but subsequently is assumed to fall back to a value close to the 28-day figure. It may be assumed therefore that the LOP remains constant at the 28-day value. The MOR falls from the 28-day value but is seen to be levelling off between 3 and 5 years and is assumed to remain approximately constant thereafter.

To maintain an adequate reserve of ultimate strength above the LOP a reduction coefficient must be applied to the LOP at 28 days. If it is assumed that this ratio should be at least 2.0, then this gives a value for D of 1.5 $\left(\text{i.e. } \frac{1}{0.65}\right)$. Again if it is assumed that sanded mixes will show some improved performance, D may be reduced to 1.4 $\left(\text{i.e. } \frac{1}{0.7}\right)$.

Handling and permanent formwork

When the material is loaded between 28 and 90 days there is little or no reduction in the LOP or the MOR:LOP ratios for both air-stored and natural weathered specimens. Thus D may be taken as 1.0. If material is loaded beyond 90 days, the factor should be separately assessed.

Mode of collapse and consequences of failure γ_c

This is the product of two factors γ_{C1} and γ_{C2}. γ_{C1} takes account of the nature of the element or structure and whether the failure mode is brittle or ductile. γ_{C2} accounts for the seriousness of collapse or failure in economic and human errors.

The use of γ_C and its incorporation into Codes of Practice for structural design is the subject of current research. In the meantime a γ_C value of 1.0 is proposed, except where there are particular circumstances involving major risks in human or economic terms, then values of 1.2 to 1.5 should be considered.

10.3.2 Other factors and considerations

Shrinkage stress, F_{ss}

As noted, there is insufficient information on the shrinkage behaviour of grc but an assessment can be made from a study of the published information for cement mortars and concretes. The conclusion reached from this is that with air-stored grc under full restraint the short-term induced shrinkage stresses will be between 1.0 and 2.0 N/mm². For natural weathering grc the shrinkage stresses should be less, say 1.0 to 1.5 N/mm². The long-term stresses will fall to about 50% of the short-term values. Where conditions of full restraint do not exist these values may be reduced.

Thermal stress F_{ts}

Again there is insufficient information on the thermal properties and behaviour of grc to calculate the thermal stresses which will occur in practice. An assessment can be made, however, based on the limited information available regarding the stresses that are likely to be induced.

F_{ts} for dry grc is likely to lie between 0.2 and 2.0 N/mm² depending on the temperature gradient at the location being considered. For damp grc F_{ts} will lie between 0.4 and 2.5 N/mm². Where conditions of full restraint do not exist, these values may be reduced.

The values quoted for F_{ss} and F_{ts} are design stresses and include the partial factor γ_c.

Box sections and sandwich panels

Infills used in forming box sections and sandwich panels may be divided into three categories:

- Non-structural
- Semi-structural
- Structural

Where non-structural infills are used the strength of the section depends only on the geometry of the grc, whereas for a structural infill full transmission of force between grc skins is assumed. Semi-structural infills lie between these extremes. Thus for rectangular sections, the effective flange width (as an L-beam) is taken as:

Infill	Effective flange width
Non-structural	$0.1\,L \leqslant 0.5\,b$
Semi-structural	$0.2\,L \leqslant 0.5\,b$
Structural	$0.5\,b$

where:
L = span of element
b = width of element

10.3.3 Summary of factors

Load increase coefficient $\gamma_f = \gamma_{f1} \times E_1 \times E_2$

where:
γ_{f1} = 0.9 to 1.6 (load conditions)
E_1 = 1.0 to 1.2 (skin thickness)
E_2 = 1.0 to 1.5 (size effect)

Strength reduction coefficient $\gamma_m = \gamma_{m1} \times D$

where:
γ_{m1} = 1.3 to 1.5 (material variability)
D = 1.0 to 1.5 (strength reduction)

Mode of collapse and consequences of failure γ_c = 1.0 to 1.5

Conditions other than bending

In general the load capacity of grc members in bending will be the critical condition but other conditions also have to be considered, viz:

- Shear
- Compression and bearing
- Deflection
- Buckling
- Fixings.

These may be dealt with in a similar manner to bending, although there is a lack of information on certain properties. In particular there is little recorded on the long-term behaviour in shear for different environments, the relationship between shear and tensile strength, and the effect of shrinkage and thermal stresses on shear. In certain cases this could be critical.

10.4 DESIGN EXAMPLE

General data

A roof unit (Figure 10.4) is required to span 2.0 m carrying a uniformly distributed load of 1.8 kN plus a self weight load of 1.4 kN. The panel is to be produced by hand spray using non-structural infill. The grc is 'standard mix' with nominal 5% fibre content. Both panel surfaces are taken to be in a dry state.

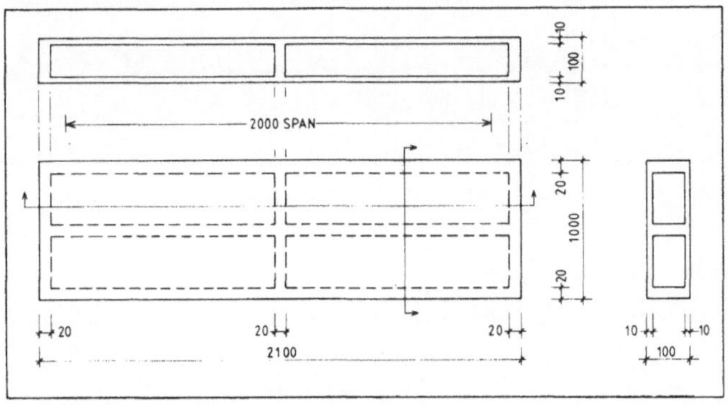

Figure 10.4 Design example—roof unit.

Section properties

Effective flange width $bo = 0.1\,L \leqslant 0.5\,b$

where:
L = 2000 mm
b = 500 mm

thus bo = 200 mm

For full unit, effective flange width = $4bo$ = 800 mm

Section modulus $Z = 0.7 \times 10^6$ mm³

Partial factors

γ_{f1} = 1.6 (live load) and 1.4 (dead load)

where:
E_1 = 1.05 (hand spray, plain panel)
E_2 = 1.25 (depth factor)
γ_c = 1.0 (mode and consequences of collapse)
γ_{m1} = 1.5 (material variability)
D = 1.3 (strength reduction, dry condition)

Loads

Design load (live) $F_{d1} = \gamma_f \times F_k$

where:
F_k = 1.8 kN (uniformly distributed)
$\gamma_f = \gamma_{f1} \times E_1 \times E_2$
$= 1.6 \times 1.05 \times 1.25$
$= 2.1$

$F_{d1} = 2.1 \times 1.8 = 3.78$ kN (live)

Design load (dead) $F_{dd} = \gamma_f \times F_k$

where:

F_k = 1.4 kN (uniformly distributed)

$\gamma_f = \gamma_{f1} \times E_1 \times E_2$

$= 1.4 \times 1.05 \times 1.25$

$= 1.84$

$F_{dd} = 1.84 \times 1.4 = 2.57$ kN (dead)

Total design load $F_d = 6.35 \times \gamma_c = 6.35$ kN

Design bending moment $M_x = 6.35 \times 2.0/8$

$= 1.59$ kNm

Design bending stress $F_x = M/z = \dfrac{1.59 \times 10^3 \times 10^3}{0.7 \times 10^6}$

$= 2.27\ \text{N/mm}^2$

Shrinkage stress $F_{ss} = 0.4\ \text{N/mm}^2$ (long term, little restraint)

Thermal stress $F_{ts} = 1.2\ \text{N/mm}^2$ (dry conditions, moderate temperature gradient)

Total design stress (long term) $= 3.87\ \text{N/mm}^2$

Load resistance

Design strength $f_d = f_k/\gamma_m$

where:

f_k = characteristic LOP at 28 days

$\gamma_m = \gamma_{m1} \times D$

$= 1.5 \times 1.3 = 1.96$

for $f_d \geqslant F_d$

$F_d \geqslant 3.87\ \text{N/mm}^2$

i.e. $f_k = \gamma_m f_d = 7.55\ \text{N/mm}^2$

Thus the grc required should have a characteristic value for LOP in bending at 28 days of say 8.0 N/mm². This would require a mean value for LOP of about 10.0 N/mm².

10.5 PERMANENT FORMWORK DESIGN

A design guide is now available, published by the GRCA (Glassfibre Reinforced Cement Association) which deals specifically with the specification and design of grc permanent formwork. The document is titled *GRC Handbook No. 1—Formwork.*

11

Quality control, testing methods and statistical evaluation of results

'Quality control' is perhaps the most misused term in the grc industry because quality is not usually controlled as such but merely recorded. Post-production recording of quality should prevent defective units leaving the factory but, if this is the only method applied, the consequence is expensive wastage. A better approach is to consider the subject as covering three definite areas:

- Pre-production checking
- Production control
- Quality recording

The choice and extent of employment of the various techniques and facilities readily available for quality control depends on the specification, the variability encountered in past production, and the decision of the management staff. A new process should, however, be handled with cautious pessimism, and then a gradual approach made towards a standard of generally acceptable cost effective quality control procedure.

Before investigating the processes of quality control, however, it is beneficial to consider the reasons why these processes are necessary and the corresponding efforts required in order to achieve the desired cost effective control over production quality. As common sense dictates, these efforts should be adequate but not excessive. It must be remembered that quality control costs are carried in the selling price of the final product.

11.1 ADVANTAGES OF STATISTICAL QUALITY CONTROL

The advantages of statistical quality control are as follows:

1. Reduction in scrap and wastage and, therefore, production costs
2. If applied during manufacture, reduction in post-manufacture inspections
3. Improvement in the morale of the work force. If production personnel are trained to understand or at least made aware of the data obtained from these techniques, they can gain a feeling of purpose and know exactly what part they play in the manufacturing process. Similarly the required standard of workmanship would be clearly defined. Personnel can then see when things go wrong and, if necessary, make the required corrections themselves.
4. Accumulation of valuable information regarding the outputs from particular machines, equipment or spray teams. It is, therefore, possible to tell whether the process can meet the specification for both current and future contracts.
5. Helpful indication of when plant maintenance is required.

11.2 THE FIVE BASIC STAGES OF QUALITY CONTROL

The five basic stages of quality control are now detailed.

Set the specification

Make what is required by the market or contract and base the specification on the particular use for which the product is intended. Consider the consequences of a sub-standard or poor quality unit leaving the factory site and being discovered later at the installation stage.

Plan to achieve the required quality

1. Plan methods of quality control in order to achieve the correct approach and sequence.
2. Plan equipment use to ensure that the facilities are available and operating correctly.
3. Obtain satisfactory raw materials and check specifications and costs to ensure the materials are suitable in relation to the specification, i.e. not inferior or unnecessarily superior for the job.
4. Select and train operatives to ensure that they are capable and concerned in producing products that meet the specification.
5. Plan inspection and shop-floor production control throughout the operations at every stage and not merely in the stockyard.

Manufacture 'right first time'

Check on defective work to ensure no incorrect machinery, plant or process is re-used (learn by one's mistakes).

Correct any quality shortcomings

Check to find the cause of any errors in materials, tools, machinery, moulding techniques or design. Provide adequate feed-back to ensure no repeat of shortcomings.

Consider long-term quality planning

Ensure the standard obtained at the works is documented for future job evaluation. This may indicate a number of significant points: the need for new plant to handle new contracts, the personnel best suited to particular tasks, and the likelihood of rejects on particular contract specifications.

11.3 PRE-PRODUCTION CHECKING

Pre-production checking must be thorough and cover all the raw materials involved, the selected moulds and specified equipment.

11.3.1 Raw materials

The following procedures should be followed with regard to each raw material employed.

Cement

Obtain cement manufacturers' test certificates and file data on cement fineness and strength tests. If production problems indicate that a particular batch of cement is causing a false set, or perhaps a flash set, the stockpile should be sampled and a check made on the setting time using the standard Vicat Test apparatus (BS 12). Troublesome stocks of cement should be put to one side. Lumpy cement should also be investigated with reference to the post-mixing state. If high shear mixers are employed it may be possible to process lumpy cement but with forced pan mixers blockages in the mortar spray head can occur.

Sand or fillers

The grading of sands and fillers should be checked regularly and compared with the limits recommended in *Chapter 5* and also with the specification offered by the supplier. It is advisable to retain a small quantity of sand of acceptable standard as a visual cross-check of what is indeed a satisfactory material.

Usually dry sand is employed but when it is intended to use wet sand the free water content should be known. The decision is then made either to mix to a workability level by adding water (plus plasticizer if required) or to add the amount of extra water necessary to maintain the water/cement ratio. Workability must always be within the stipulated range of the manufacturing process. Ideally, batching by using whole bags of sand is often quicker and easier than weighing out additions or gauging by volume.

The above also applies in deciding whether to use wet or dry sand. If scales are used, then a simple calibration check should be performed with every shift.

Admixtures

These must be clearly labelled and not transferred into unlabelled storage jars. Additions must be made to comply with the manufacturer's recommended dosage, or to an acceptable pre-production test level. The method of addition should be as sophisticated as possible, for example by using the proper admixture dispersing equipment which is obtainable from the admixture supplier. It should be remembered that over- and under-dosing can be equally disastrous.

Water

Water should be added via gauging equipment which removes human error resulting from guess work (see *Chapter 3*). All dispensing equipment should be checked by an appointed person on a regular basis and a tabulated record kept of these checks together with the results obtained.

Fibre

Before production commences at the beginning of a shift the roving tex weight has to be noted. Production control procedures require this running check for constant monitoring of the fibre content of the product.

11.3.2 Moulds

The effective use of a mould is dependent on its cleanliness, freedom from damage and accuracy of dimensions — features which are now considered in more detail.

Cleanliness

Adherence to normal 'housekeeping' must be carried out but there are specific problems which should be drawn out for consideration.

Prior to use moulds require checking to ensure that they are not contaminated with overspray from adjacent production. This is especially important when a range of pigmented or white and ordinary Portland cement products is passing through the factory. Even if only one type of matrix is involved, inclusion of overspray into the surface can produce a 'grainy' porous texture that is patchy and may occur in local areas on the surface. A simple precaution is to cover the moulds with sheeting immediately after applying the mould oil. This will also protect against general dust and spillage.

Grc moulding profiles are usually of such a shape as to fully exploit the 'mouldability' of the material. Hence the moulds are often designed so that intricate detailing is produced by removable sections, resulting in profiles of close tolerance which may be even more fine than those experienced in normal precast concrete production. This means that build-ups of excess composite must be removed before they harden and lead to dimensional errors which result in excessive time having to be spent on demoulding and re-assembly.

Finally, mould oil application should be carefully controlled to prevent 'ponding' and over-application of mould oil in corners and such like, which produce soft crumbling edges on demoulding plus staining and inconsistent colour over the mould face.

Damage

Moulds require constant checking to ensure that any damage that might have occurred is noted and rectified. It is always good practice to over-strengthen moulds where handling by fork-lift trucks is envisaged because of the greater likelihood of impact damage. A system of reporting noted damage is also advisable in addition to recording ageing and breakage details when wooden moulds are used.

Dimensions

Mould checks after re-assembly must not be made with tape measures but with jigs. Length checks should be carried out using pre-cut lengths of wood or polypropylene taken from the mould shop; in fact the moulds need to be assembled with these as guides. The squareness should also be checked with a pre-cut triangle manufactured in the mould shop when the first mould was being completed, taking the inspected mould as the master. The location of any feature line formed in a panel needs to be made with dowels and checked with pre-cut guide lengths of timber, steel or plastic.

11.3.3 Equipment

Before any grc is sprayed at the start of the shift, the mortar output and the chopped glassfibre output need to be checked and probably reset to ensure that the correct percentage fibre content is achieved. The two outputs are monitored separately at the side of the first mould in the production line. The procedure is as follows.

Determination of the delivery rate of a glassfibre chopper
A weighed polythene bag (W_1) is held under the outlet of the chopper gun so that all the chopped fibre output can be collected. At one of the bottom corners of the polythene bag a 10 mm diagonal cut is made to allow the air from the air mover to pass out of the bag preventing back pressure building up and blocking the path of the escaping chopped fibre. This method enables the neck of the polythene bag to be tightly clamped, by hand, around the output nozzle of the chopper gun. The glass chopped in a 15 s running period is collected and the bag and its contents weighed (W_2). Three such checks are usually made and the mean is calculated in g/min.

Delivery rate = $4(W_2 - W_1)$ g/min.

Determination of the delivery rate of cement slurry from a spray head
A weighed bucket (W_3) is positioned under the mortar spray nozzle and the slurry collected for a spray period of 30 s under actual operating conditions. The bucket and its contents are weighed (W_4). The mean of three such measurements is made and the delivery rate is calculated in kg/min. The slurry is then returned to the pump hopper.

Delivery rate = $2(W_4 - W_3)$ kg/min.

The two outputs obtained are then combined to achieve the setting for the percentage fibre inclusion.

Combination of results to obtain percentage fibre inclusion
The percentage fibre inclusion is expressed as the percentage of fibre relative to the total wet solids (including fibre), i.e.

$$\%\text{ fibre inclusion} = \frac{\text{Fibre inclusion} \times 100}{\text{Mortar} + \text{Fibre inclusion}}$$

thus:

$$\%\text{ fibre inclusion} = \frac{(W_2 - W_1)}{(W_2 - W_1) + 500(W_4 - W_3)}$$

where:

$\left.\begin{matrix} W_1 \\ W_2 \end{matrix}\right\}$ in grammes

$\left.\begin{matrix} W_3 \\ W_4 \end{matrix}\right\}$ in kilogrammes

Spray pattern setting
Once the spray equipment has been set to produce the required fibre content, the spray pattern from the spray head has to be checked to ensure that even distribution of the fibre in the mortar cone is achieved. The specified composite should then exhibit even distribution of uniformly coated fibres without leaving plain mortar or uncoated fibre behind. The effect of any maladjustment is to produce 'veins' of dry glass within the composite cross-section, leading to a drastic drop off in inter-lamina shear strength. In fact test coupons cut from the material, when exhibited to the customary four-point bending test described later, may fail by delamination rather than in the customary flexural mode. This is shown in Figure 11.1. The sequence of setting-up the

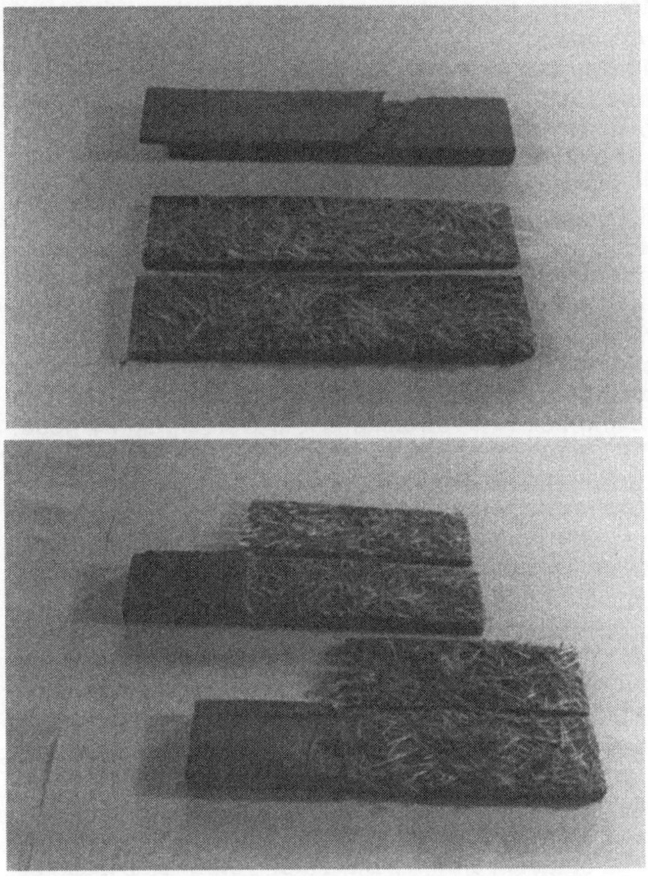

Figure 11.1 Delamination failure of grc test coupons with dry glass veins in the cross section. This results from incorrect setting of the equipment.

required spray lay-down pattern is described in *Chapter 7*.

11.4 PRODUCTION CONTROL
Production control is achieved by the methods now outlined.

11.4.1 Continuous monitoring of fibre content
In order that fibre content of the sprayed composite is maintained at an acceptable level, continuous monitoring is vital. This can be achieved by the following means.

Monitoring the fibre throughput as a weight loss from the box of rovings
This method is the simplest means of determining the fibre content while production is underway. The box of rovings is located on a balance and the registered weight adjusted to read at a convenient starting point (i.e. the nearest whole kg marker). On knowing the output from the mortar spray gun, determined from the calibration test, the required inclusion of rovings can be deduced for the intended fibre content. This would be expressed as a weight of rovings chopped and sprayed in 1 min. Hence, if the required fibre usage is for example 600 g/min., the amount of fibre going through the process can be found by noting the indicated weight change during a 1 min. interval. This monitoring may be applied every 30 min., or whenever there is considered to have been a change in the fibre content.

Monitoring the mortar pump output and the roving take-up speed

This technique has the distinct advantage of being a true continuous check on fibre content with the possibility of adaptation for self correction to maintain the required fibre content.

The output from a grc mortar pump is proportional to the speed of the pump (see Figure 11.2). Consequently, if the speed of the pump shaft is measured the output from the pump can be determined and checked. Similarly, the quantity of glassfibre being processed can be monitored through the measurement of the roving speed. This is achieved by causing the roving to drive a rubber-faced slave wheel situated just 1 m above the roving box (with an automatic traversing spray head, the slave wheel has to be located on the traversing carriage). Knowing the diameter of the slave wheel and the speed of rotation, the fibre throughput can be determined. This method accounts for any slippage that might occur at the chopper gun; merely measuring chopper gun speed is not reliable under operating conditions (see Figure 11.3). To summarize, knowing mortar output as a function of the pump speed and fibre throughput as a function of the roving slave wheel speed, the composite fibre content can be monitored in these terms.

The graph shown in Figure 11.3 illustrates the method of determining fibre content. Mortar pump output in kg/min. is plotted along the x axis with the glassfibre output in g/min. plotted along the y axis. It can be seen that by the method of defining fibre content, i.e.,

$$\% \text{ fibre content} = \frac{\text{Fibre output}}{\text{Fibre output} + \text{Mortar output}}$$

each particular percentage level of fibre content can be

Figure 11.2 Calibration of the output from an MT 420 Mono Pump.

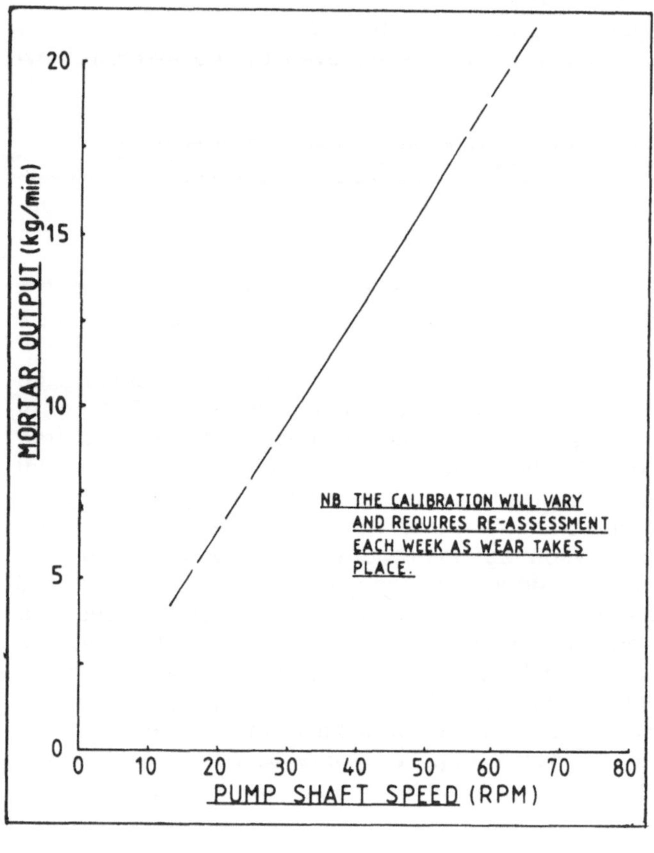

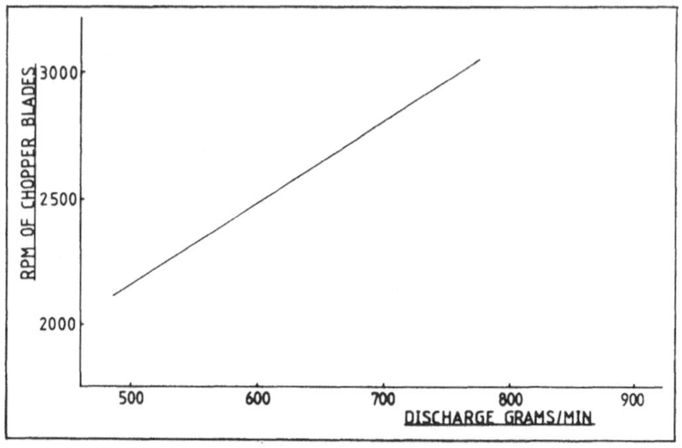

Figure 11.3 Discharge from Downland chopper with air mover.

given as a ratio which is represented by the slope of its line on the graph. The determination of the fibre mortar relationship can be explained further where:

X = glassfibre weight (kg)
Mt = mortar weight (kg)
n = % fibre content of the composite

By definition:

$$\% \text{ fibre content} = \frac{\text{Fibre weight}}{\text{Fibre weight} + \text{Mortar weight}}$$

hence:

$$\frac{n}{100} = \frac{X}{X + Mt}$$

and by arrangement:

$$Mt = X\left(\frac{100}{n}\right) - 1$$

now:

$$Mt = X.C \text{ where } C = \left(\frac{100}{n}\right) - 1$$

or

$$X = \frac{Mt}{C}$$

$$= Mt.\left(\frac{1}{C}\right)$$

Thus, by selecting the appropriate value of C or $\frac{1}{C}$ from Table 11.1 for the relevant percentage fibre content, it is possible to determine the ratio of mortar:glassfibre. This naturally applies to all rates of output. For example at 5% fibre content, from Table 11.1, 1 kg of glassfibre requires 19 kg of mortar.

By calibrating the mortar pump output and inserting the calibration values along the x axis in the spaces provided, the Table is complete. The six vertical columns of roving slave wheel speeds are obtained by using a 57.5 mm diameter slave wheel (a conventional packing roller) and by taking six particular values for roving tex weights, i.e., the weight of a 1 m length of fibre. This is normally 2.45 g/m and definitely within the range 2.45–2.20 g/m.

Hence, knowing the required fibre content level, the graph yields the quantities of fibre and mortar. The rate of output has then to be fixed and the corresponding fibre and mortar quantities read. This will give a

Table 11.1 Table of ratios of fibreglass: mortar weights for particular % fibre contents.

% Fibre content	C	$1/C.10^{-2}$	% Fibre content	C	$1/C.10^{-2}$
2.0	49.00	2.040	5.1	18.61	5.374
2.1	46.62	2.145	5.2	18.23	5.485
2.2	44.45	2.249	5.3	17.87	5.597
2.3	42.48	2.354	5.4	17.52	5.708
2.4	40.67	2.459	5.5	17.18	5.820
2.5	39.00	2.564	5.6	16.86	5.932
2.6	37.46	2.669	5.7	16.54	6.045
2.7	36.04	2.775	5.8	16.24	6.157
2.8	34.71	2.881	5.9	15.95	6.270
2.9	33.48	2.987	6.0	15.67	6.383
3.0	32.33	3.093	6.1	15.39	6.496
3.1	31.26	3.199	6.2	15.13	6.609
3.2	30.25	3.306	6.3	14.87	6.724
3.3	29.30	3.413	6.4	14.63	6.838
3.4	28.41	3.520	6.5	14.38	6.952
3.5	27.57	3.627	6.6	14.15	7.067
3.6	26.78	3.735	6.7	13.93	7.181
3.7	26.03	3.842	6.8	13.71	7.297
3.8	25.32	3.950	6.9	13.49	7.412
3.9	24.64	4.058	7.0	13.28	7.527
4.0	24.00	4.167	7.1	13.08	7.643
4.1	23.39	4.275	7.2	12.89	7.759
4.2	22.81	4.384	7.3	12.70	7.875
4.3	22.26	4.493	7.4	12.51	7.992
4.4	21.73	4.603	7.5	12.33	8.108
4.5	21.22	4.712	7.6	12.16	8.226
4.6	20.74	4.822	7.7	11.99	8.342
4.7	20.28	4.932	7.8	11.82	8.460
4.8	19.83	5.042	7.9	11.66	8.578
4.9	19.41	5.153	8.0	11.50	8.696
5.0	19.00	5.263	—	—	—

corresponding speed setting for the mortar pump and for the fibre slave wheel, the latter being dependent on the particular tex weight of the roving.

Finally, it is possible to take this a stage further. Since the speed measurements are preferably made with proximity transducers working on the magnetic field principle, the outputs of the transducers can be fed to a ratiometer. This meter can then take the mortar output (expressed as pump rpm) as the base and process the fibre usage (expressed as slave wheel rpm) as a ratio of the mortar output expressing the principle shown in Table 11.1. Information may be displayed on a dial indicator or pen recorder. Limits in the system can be set to correct the chopper gun output ensuring that the fibre content is within a prescribed tolerance band. The mechanics of this self correction system are better suited to a chopped gun driven by an electric motor. The consequent extra weight that would be added to the spray head favours an automatic traversing set-up.

Assuming that both the chopper gun and the mortar pump are driven by 3-phase a.c. motors of 220 V then, by using synchronous motors, they can be driven via controllers to maintain constant torque for varying output speeds. Also, since the speed of a synchronous motor is a function of the input frequency, the speed can readily be varied and maintained. Controllers are available that will accept 220–240 V single phase (50–60 Hz) normal mains supply and drive a suitably adapted 3-phase synchronous motor to maintain a set speed to $\pm 0.5\%$. Other types will control two or more motors to maintain a set speed ratio. In this 'ratio setting' mode, the 'demand' input is made common to both motor drive modules. The 'analogue gearbox' output of the second module (driving the chopper gun) is brought out to a potentiometer, setting a ratio against the master speed demand (for the mortar pump). With this connection, both the speed controllers accelerate and decelerate the drive motors together while the set ratios are maintained, the ratios being chosen from the data in Figure 11.4.

The current cost of this type of system may be around £1500 but the advantages can be expressed as follows:

- Fibre content can be maintained with confidence to the specified level
- With precisely maintained fibre content, lower levels can be set with confidence especially for products such as permanent formwork, where lower amounts can be tolerated
- The cost of allowing for variations in fibre content, namely setting at 5.5% to guarantee 5.0%, is removed from the production cost of the products.

Monitoring by weight loss

A further development of the system described in the previous section is now under investigation. The same principle of continuous monitoring of mortar and fibre usage is employed but the method of measuring is changed to focus on the weight loss.

The mortar pump is suspended via cables from an overhead gantry, with a load cell incorporated into the suspension system. By this means the load cell can be arranged to record weight loss as the pump delivers mortar to the sprayhead and subsequently onto the mould. Similarly, the roving cheese is positioned on a load cell and, again, weight loss of the roving is measured as fibre is incorporated into the composite.

Obviously these weight losses are measurements of material usage and the output from the load cells can be integrated to give a ratio of fibre to mortar, with the mortar usage as the base (column C in Table 11.1). Again, it must be remembered that fibre content is expressed as $\frac{\text{Fibre} \times 100}{\text{Fibre} + \text{Mortar}}\%$.

The ratios given in column C of Table 11.1 are expressed in this form. Consequently 5% fibre inclusion is denoted as a ratio of 19, that is 19 parts mortar and 1 part fibreglass (both by weight) — not 20 parts mortar as might be expected.

The final feature of this system is built in feedback controllers. These will keep the motors that drive the mortar pump and chopper gun in sequence so that the required fibre content is continuously maintained within a tolerance band.

It is suggested that the mortar pump output is set first. The chopper gun output can then be set to match. The system is designed to measure the weight losses (and hence the material usage) of both the pump and roving cheese for a 15 s interval, and to integrate these results. A correction would be instituted if the required ratio was not observed (within the limits of the tolerance). This should be brought into the next 15 s phase.

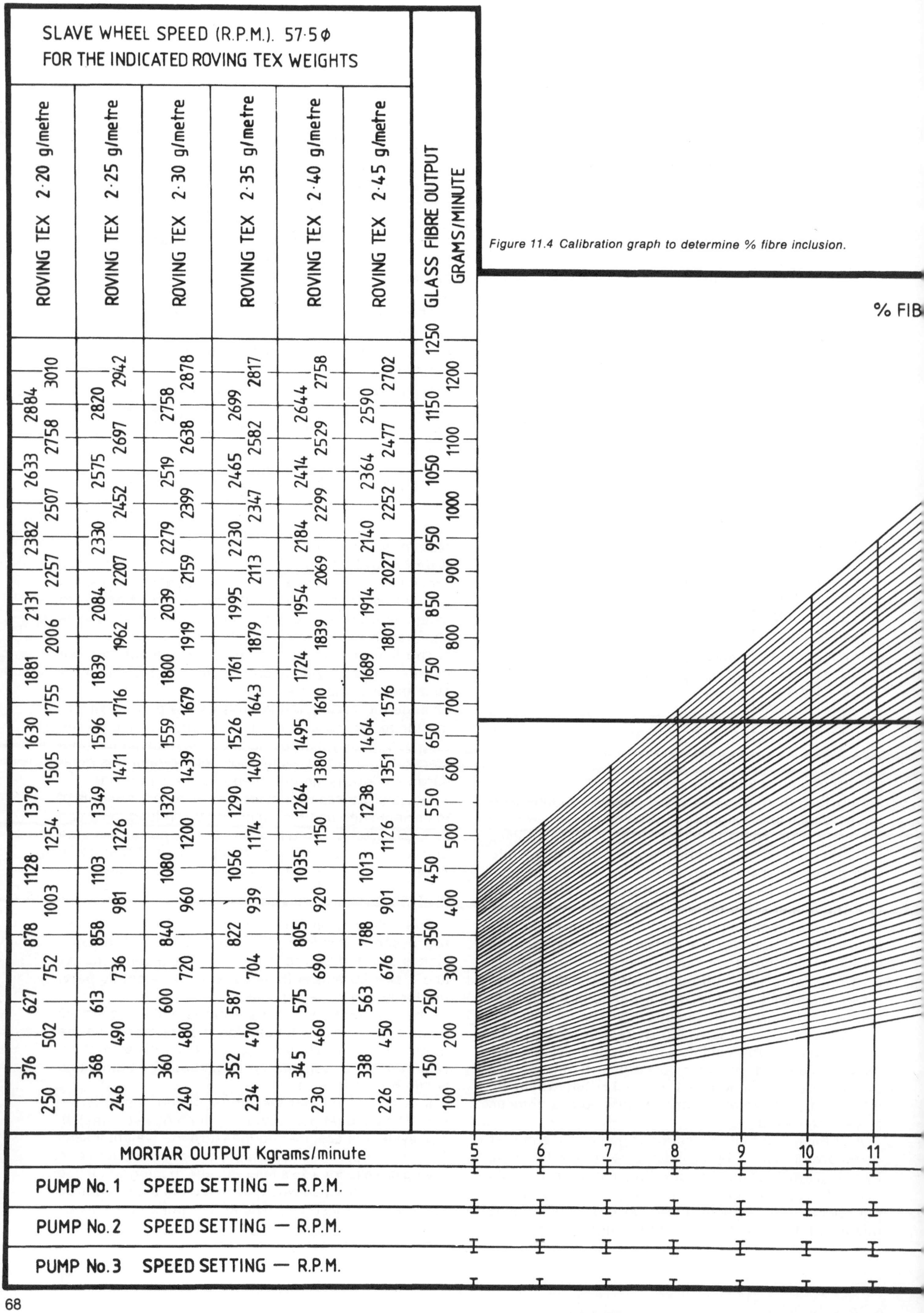

Figure 11.4 Calibration graph to determine % fibre inclusion.

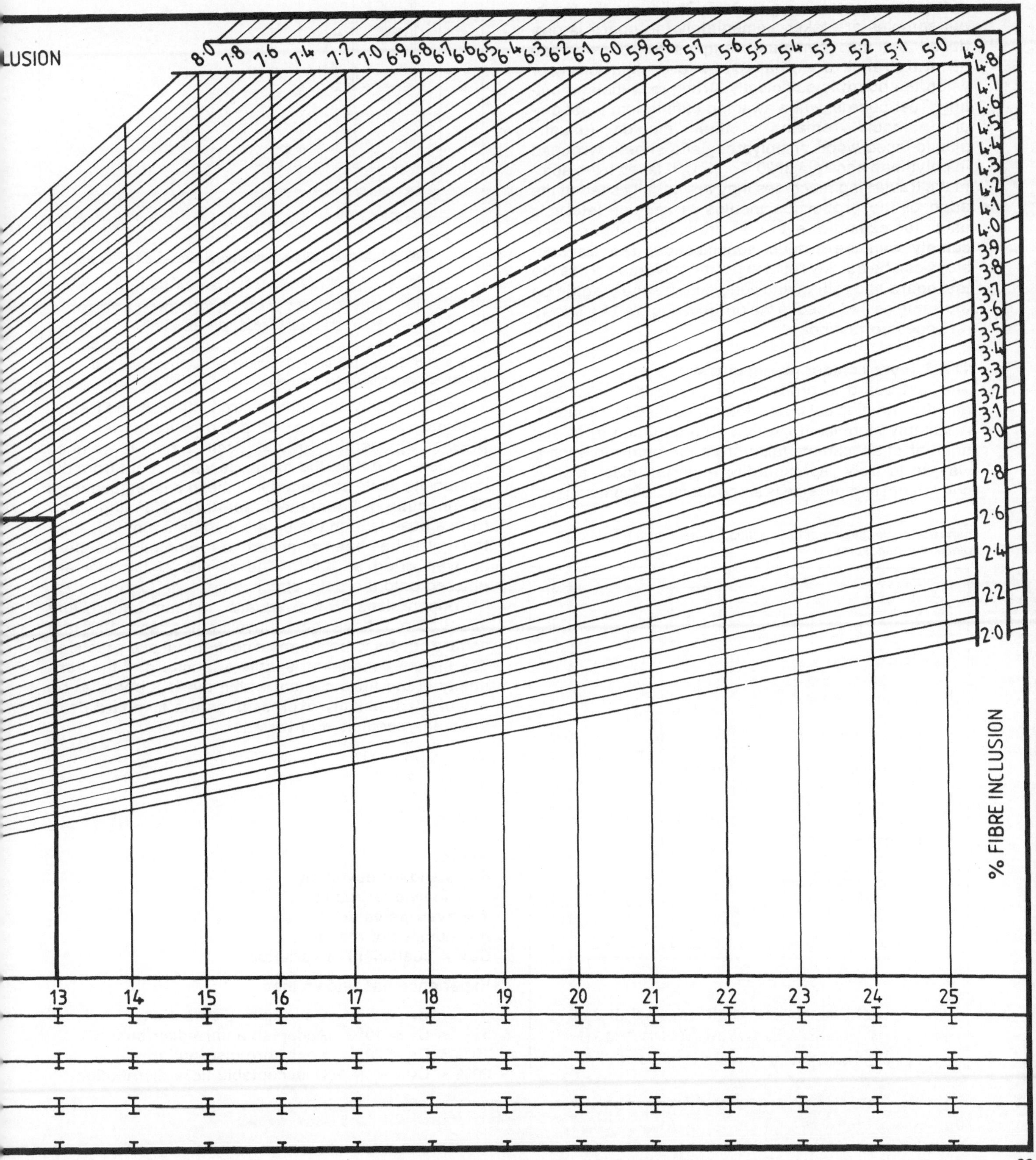

LUSION
8·0 7·8 7·6 7·4 7·2 7·0 6·9 6·8 6·7 6·6 6·5 6·4 6·3 6·2 6·1 6·0 5·9 5·8 5·7 5·6 5·5 5·4 5·3 5·2 5·1 5·0 4·9
4·8 4·7 4·6 4·5 4·4 4·3 4·2 4·1 4·0 3·9 3·8 3·7 3·6 3·5 3·4 3·3 3·2 3·1 3·0 2·8 2·6 2·4 2·2 2·0
% FIBRE INCLUSION
13 14 15 16 17 18 19 20 21 22 23 24 25

The cost of this complete controlling and recording system is estimated at £7000 at today's prices.

11.4.2 Manufacture of test boards

During the production of grc units a test board is sprayed. Samples are taken from here to determine the fibre content (washout test), the density and the flexural strength at both 7 and 28 days. The usual size for the test board is 600 mm square and the thickness is kept to a 10 mm maximum. The test board is produced by the same manner of operation and receives the same degree of roller compaction as the production units. A suggested plan for cutting the board into test pieces is indicated in Figure 11.5. Obviously, the washout samples are taken immediately the board is produced but the remaining samples for density and flexural strength are required after overnight curing, or when the board is sufficiently strong to withstand cutting. Test boards, while awaiting cutting into test samples, should remain in their moulds (usually plastic-faced wood or polypropylene sheet), wrapped in polythene and kept at 20°C ± 5°C. If taken from automatic traversing machines they are usually sited at the edge of the production moulds in regular intervals after, for example, every 30 production outputs and comply with the product specification. The traverse stops can be quickly reset to permit traverse across the test board and rolling and compaction should again be the same as that used in the product. Figure 11.6 shows a suggested test board.

11.4.3 Fibre content (washout tests)

The test board is produced as previously described and the samples cut out to determine the fibre content. A suitably significant result is that obtained from any number of specimens taken from the board in such a way as to truly represent the material under test. A convenient sample size is six coupons, 100 mm square.

Figure 11.5 Suggested plan for cutting a test board into test pieces for handspray.

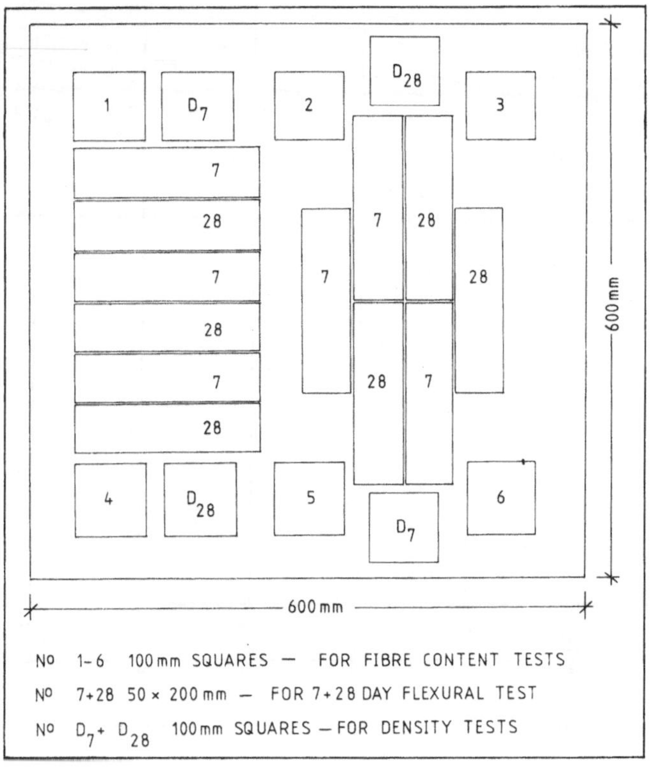

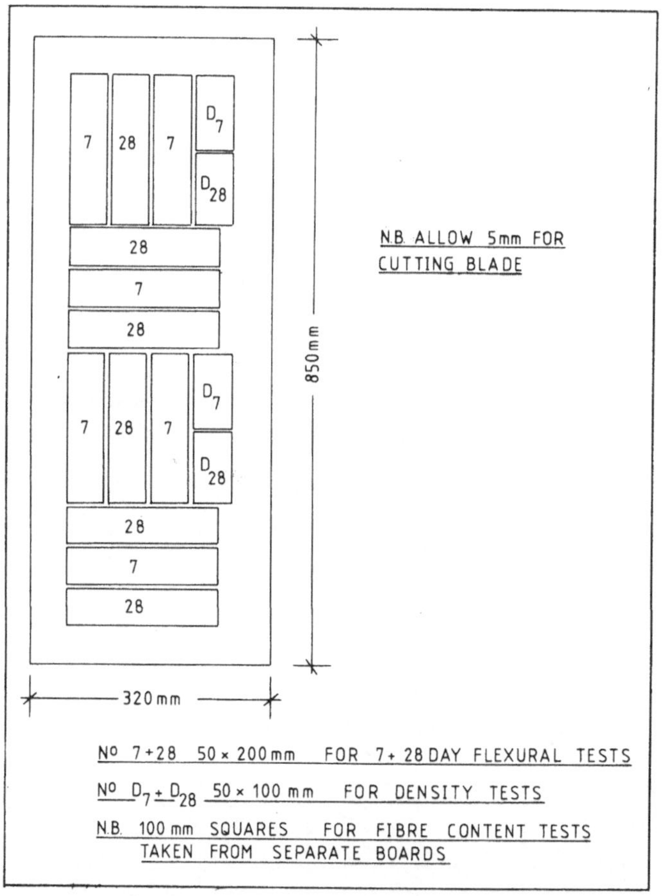

Figure 11.6 Suggested plan for cutting a test board into test pieces for automatic machine work.

These are easily cut with a sharp knife from the freshly sprayed board using a polypropylene template pressed against the surface. Each sample is weighed complete, and then the mortar is washed out from the composite while contained in a mesh basket (usually a 75 mm cube of 2 mm mesh aperture). Running water from a hose is found to be a convenient method. After the matrix has been removed, the remaining wet fibre is oven dried at 105°C. The results of the test are entered onto the daily check sheet (see Figure 11.7).

Coupon sizes 75 mm square, 100 mm square and 150 × 50 mm have been investigated and 100 mm square seems to be the most convenient and practical. The six coupons taken enable a measure of the variability of the fibre content within the test board to be established, usually expressed as the standard deviation and coefficient of variation.

$$S = \sqrt{\frac{\sum(x_1 - \bar{x})^2}{n - 1}}$$

$$CoV = \frac{S}{\bar{x}}$$

where:
S = standard deviation
x_1 = individual result
$\bar{x}$ = average value
n = number of results
CoV = coefficient of variation

Experience has shown that:

$Cv < 5\%$	Good fibre distribution
$5\% < Cv < 10\%$	Acceptable fibre distribution
$10\% < Cv < 20\%$	Poor fibre distribution
$20\% < Cv$	Unacceptable fibre distribution

DAILY CHECK SHEET

DATE:

CONTRACT:

BASK. No.						
BASK + SAMP Wt. BASK						
Wt. SAMP.						
BASK. + GLASS Wt. BASK.						
Wt. GLASS						
% GLASS						

Average (6)

Check Made on Glass Rovings = ______ gms/metres length

Output:- Glass = ______ kg/min

Slurry = ______ kg/ min

% According to bag and bucket test = ______ %

% According to sample test = ______ %

Standard Deviation ______ N/mm^2

$\pm$ ______ % According to setting

Coefficient of Variation ______ %

Figure 11.7 Daily check sheet.

11.4.4 Roving weight

Before production is started, especially if the ratio method of continuous fibre content monitoring is employed, the roving weight has to be obtained. Knowing the weight enables a check to be made on the fibre quality for compliance with the specification.

The method involves recording the weight in grams of a 1 m length of the roving, making sure that the complete roving is sampled. The weight should be 2.50 g for the metre length. This can also be expressed in tex form as 2500 in g/km.

At any stage during production when the fibre content is either visually assessed or monitored as being low, the roving weight should be determined. It is rarely found to fall below 10% of the mean value, but continuous monitoring will enable the current situation to be revealed. However, a 10% drop in roving weight is registered as a 0.5% change in fibre content. By monitoring the fibre weight variations, therefore, the recorded standard deviation can be employed enabling an oversetting fibre content to be made to ensure that, with drops in fibre weight (down end rovings), the specification is met.

11.4.5 Water/solids ratio of the matrix

This is determined by drying in a muffle furnace at 500°C for 30 min. noting the weight before and after drying (the sample or samples being taken from the test board).

$$\text{Water/solids ratio} = \frac{(W_2 - W_1) - (W_3 - W_1)}{(W_3 - W_1) - \frac{G}{100}(W_2 - W_1)}$$

where:
G = average glass content
W_1 = weight of basket
W_2 = weight of basket + specimen
W_3 = weight of basket + dry specimen

If additions of latex admixtures or materials liable to decompose at 500°C are included in the matrix, the result will naturally be affected.

An alternative form of quick and effective measurement of moisture content can be made with a calibrated siphon can (as used for determining the moisture content of sand). Correction for the fibre inclusion would, however, have to be made.

11.4.6 Fresh wet density

This property is of vital importance in determining the quantities of materials required to produce a particular unit. Practical experience has found that the best method is to spray-up and roll, again in the typical production manner, a mould of 300 mm × 300 mm × 10 mm thickness (or whatever thickness is intended for the product). The back face is then struck off flush with the mould sides and the weight noted.

$$F.W.D = \frac{W_2 - W_1}{9 \times 10^{-4}} \text{kg/m}^3$$

where:
W_2 = weight of mould kg + sample
W_1 = weight of mould kg

The mould should be manufactured from a water resistant material, preferably polypropylene or a similar plastic. After the weighing, the mould should be emptied and washed thoroughly.

11.4.7 Measurement of the consistency of a matrix material

As with all the common methods of measuring the consistency of a cementitious material (namely slump, VeBe and compaction factor) usually single point tests are performed at one rate of shear but these do not define the complete relationship of shear stress α rate of shear. The 'Tattersall' method, however, enables a more true picture to be obtained of the rheology of the material. For a simple test, therefore, on a matrix containing no admixtures, the flow table type test may be satisfactory, but when stiffening admixtures are employed to enable vertical spraying to be performed, the Tattersall method is preferred. This method can be used to indicate changes in the water/cement ratios for matrices of fixed solids proportions. Consequently, the ability to define a yield stress and plastic viscosity is of great benefit, once the parameters of the spraying and manufacturing system have been defined and especially when rapid setting cements (Schnell cement) and consistency modifying admixtures are to be employed.

11.4.8 Product thickness checks

Various checks can be carried out to ensure that the correct product thickness has been achieved.

Pin gauges

The thickness of a sprayed layer of composite can simply be checked with a calibrated rod pushed in down to the mould face. This enables the production team, and especially the sprayer, to assess the required thickness necessary to bring the panel up to the required specification.

Indentation test gauges

A further derivation of the pin gauge is the indentation

gauge which has a wire cross or circle arrangement attached at right angles to the pin shaft. This produces a pattern on the grc surface if the composite is equal to, or greater than, the required thickness. Its chief benefits are that it offers the production staff a simple visual check and acts as a permanent mark which shows that the thickness has been achieved (for final inspection). Production thickness limits can be checked by having two patterns, for example a cross for minimum thickness and a circle for maximum thickness. This method is mainly used by manufacturers producing permanent formwork units or products where the visual appearance of the back of the unit is not critical. A probe is shown in Figures 11.8 and 11.9.

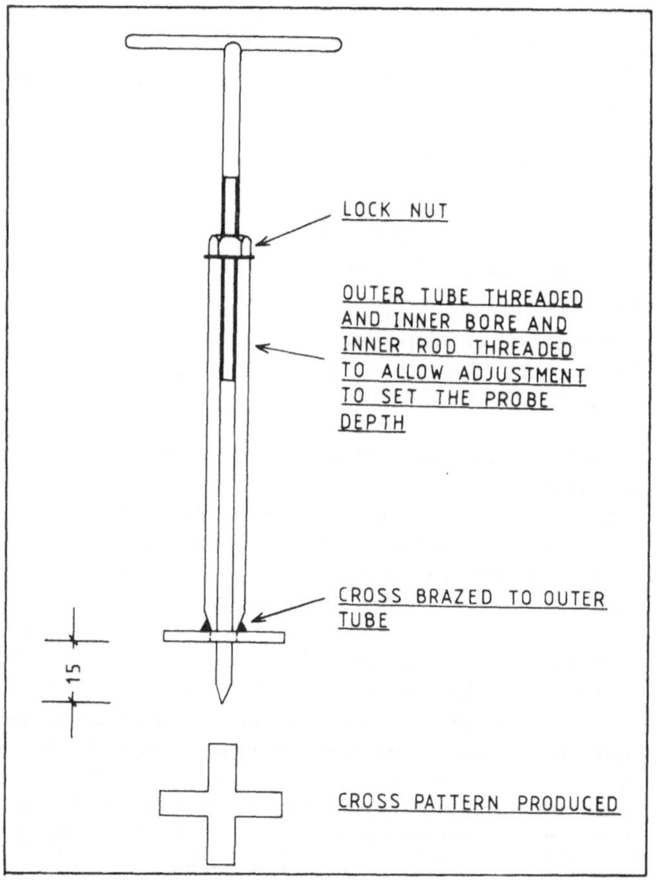

Figure 11.8 Indentation probe.

Figure 11.9 Checking depth of composite lay-up using cross head probe.

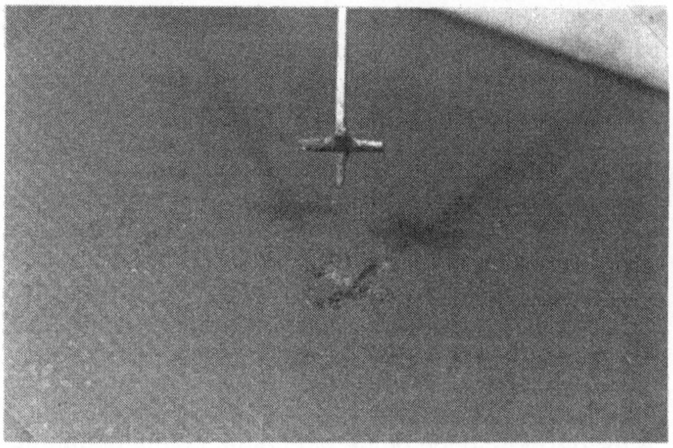

Covermeter method

If production is carried out using steel moulds, steel faced moulds or wooden moulds with steel inserts or backing plates at important areas, thickness can be measured using these devices. Figure 11.10 shows a calibration graph produced to check single skin material using steel moulds ranging in thickness from 18→3 swg (1.3–6.5 mm). The device was calibrated on the moulds using blocks of polypropylene, before the material was sprayed. This system could be adapted to produce a production check for hand spray work and also, with the addition of a feedback loop, to control the traverse rate of the cross on an automatic spray machine for uniform thickness.

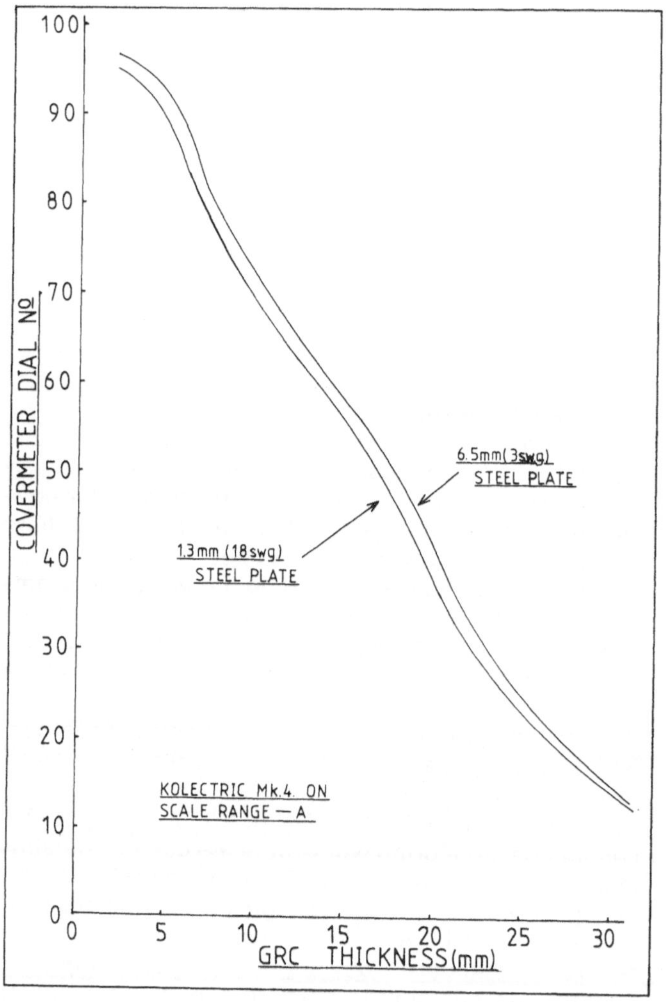

Figure 11.10 Calibration of a covermeter to indicate composite thickness.

Final check

As a check in the stockyard or elsewhere, the product thickness can be measured after calibration by offering the sensor head up to the unit directly opposite a steel plate and then moving it to produce the maximum reading. The smoothness of the hardened panel is critical, and allowance has to be made for a lumpy surface. This lumpiness is measured with a depth micrometer.

11.5 QUALITY RECORDING

The process of quality recording is carried out on the cured test coupons cut from the test board. The objective is to obtain a representative sample of the output of the grc for the day in question and, after suitable curing and ageing, to obtain values for the physical properties of the production output by testing the sample coupons. Naturally the curing procedure followed for the test coupons will affect the properties obtained. The two basic concepts adopted are:

- To cure under water at 20 °C ± 2 °C for the duration and to test immediately on removal from the water
- To cure in the same conditions as the production output.

The latter is aimed at producing properties in the test coupons that represent the properties achieved in production. This assumption is liable to be incorrect. Curing the samples by the first method will produce a consistent, controlled environment for the test pieces and hence any property variation will be a function of the composition at the time of spray-up. This consistent curing regime also enables property changes to be compared throughout the year, unlike the variability that can occur by using the second method.

Various other curing procedures for test coupons have included:

- Under water at 20°C after 24 hours at ambient temperature while wrapped in polythene
- In air at ambient temperature
- In air at laboratory temperature (20°C)
- In air alongside the production units
- Wrapped in polythene containing 'free water' at ambient temperature
- With the production unit wrapped in polythene containing 'free water'.

11.5.1 Density

For density testing two methods are employed, namely water displacement and volume measurement, combined with weighing. Wet or oven dried samples are suitable. The method preferred by the author is to use the actual flexural test coupons prior to the strength tests. Volume is measured by water displacement employing a suitable hanging sample carrier attached to the underside of the balance. Density is found by the following:

$$D = \frac{W_1}{W_1 - W_2} \times 1000$$

where:

D = density (kg/m)
W_1 = weight in air (kg)
W_2 = weight in water (kg)

Wet density measurement of test coupons enables a direct comparison to be made between test coupon density and flexural strength, or whatever subsequent physical test is carried out. Dry density may, however, be required in design to establish product weights and so forth. This can be obtained by water displacement to determine the volume, followed by drying to a constant weight in an oven at 100 °C and then weighing.

$$\text{Dry density} = \frac{W_3}{W_2 - W_1} \times 1000$$

where:

W_3 = oven drying weight (kg)
W_1 = wet weight in air (kg)
W_2 = wet weight in water (kg)

The value of a density check in routine quality recording is to help explain low or varied results from physical property tests.

Simply measuring the volume of test pieces by using vernier calipers or similar is not satisfactory due to the uneven nature of the back face together with the non-uniform shape obtained after the test piece is removed.

11.5.2 Tensile strength

Although this property can be relatively easily measured using a variety of conventional testing equipment, the results obtained greatly depend on the preparation of the samples and the mode of operation of the clamps. However, the technique employed is to use 150 mm × 25 mm samples over a 100 mm gauge length. Wedge action or screw clamp grips ensure that the sample is aligned with the axis of the loading system. Machine cross-head speeds of 1.5–2.0 mm/min. are customary.

Most test equipment produces a load/deflection plot by either manual or automatic means. The typical trace produced from a tensile test (see Figure 11.11) yields a two-stage failure. The first region shows pseudo-elastic behaviour of a relatively high modulus limited by the cracking of the matrix. The second region exhibits progressive multiple cracking at a lower modulus ending in the enlargement of one crack and the final break up of the sample. The point of modulus change and the stage when the matrix cracks is defined as the BOP, bend-over point, and the final load indicates the UTS, ultimate tensile strength.

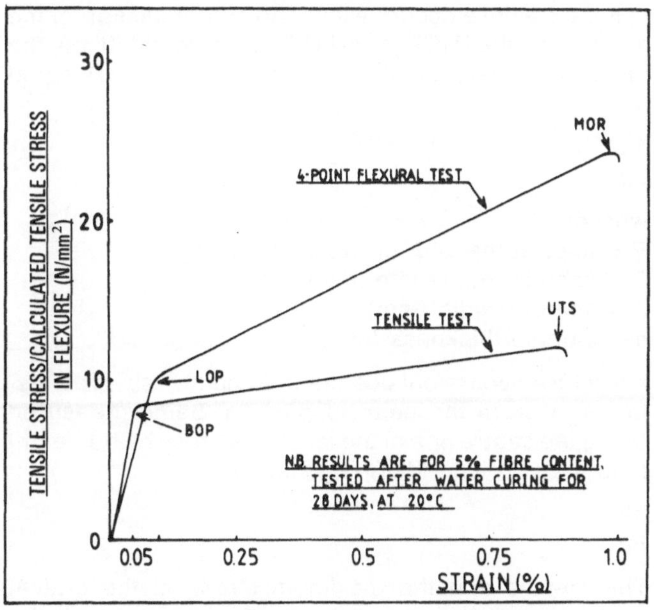

Figure 11.11 Typical tensile test and flexural test graphs.

From the graph trace and measurements of the broken test pieces (using vernier calipers), the stress at BOP and UTS is found as follows:

$$\text{Stress at BOP and UTS} = \frac{F}{BD}$$

where:

F = load at the BOP or UTS
BD = cross-sectional area of specimen.

Due to the care and precise nature required to perform this test, it renders its use more akin to research studies than quality recording procedures. Suitable equipment is, however, illustrated in the section on test equipment which appears further on in this Chapter.

11.5.3 Flexural strength

The flexural strength of grc test coupons is expressed as the LOP, limit of proportionality, and the MOR, modulus of rupture. The LOP is determined from the point on the load/deformation plot where the initial pseudo-elastic response deviates from the linear path. This coincides with the first crack developing in the matrix. The MOR is determined from the peak value in the load/deformation plot (Figure 11.11).

In general, six coupons are tested at 7 and 28 days. This may be reduced to four samples for on-going quality recording. The sample is split to yield equal numbers of coupons with the mould face in tension and compression. A further division may have to be made to check for variations in the longitudinal and transverse directions due to non-uniform fibre dispersion.

The standard test sample length is 200 mm × 50 mm × 8–10 mm thick. The four point test is the recognized method of flexural testing. This is due to the preferable situation of subjecting a zone — namely the region of the sample between the inner supports — to a uniform test in contrast to the three-point test which subjects only the plane immediately above the centre support to the maximum stress. It will therefore yield a lower value of flexural strength and higher value of tensile strength than that obtained by pure direct tension testing.

If the fracture occurs within the centre section of the specimen, the MOR and LOP are calculated on the basis of ordinary elastic theory and are therefore equal to:

$$\frac{PL}{bd^2}$$

where:
P = load at the LOP or MOR (Newtons)
L = sample test length (mm)
b = sample width (mm)
d = sample thickness (mm)

If fracture occurs outside the load points, say at a distance 'a' from the near support ('a' being measured along the centre line of the tension surface of the beam) then the LOP and MOR is given by:

$$\frac{3Pa}{bd^2}$$

This means that the maximum stress at the critical section is used and not the maximum stress on the beam. However if $\frac{L}{3} - a > 0.05L$ then the sample and result should be disregarded. Often a failure outside the central zone is due to an over-thick sample being tested, or one with a poor interlaminar shear strength.

In this test the minor span is taken as one third of the major span namely $\frac{200}{3}$ mm but, in order that the sample span to thickness ratio is retained within 16:1 and 20:1, the sample sizes as given in Table 11.2 apply. The rate of crosshead speed is kept between 1.5 – 2.0 mm/min. The sample sizes, width and thickness at the fracture point are measured with vernier calipers to 0.05 mm. The arithmetical mean, standard deviation and coefficient of variation are determined and also the characteristic LOP and MOR at both 7 and 28 days.

Table 11.2 Test sample sizes.

Thickness of specimen (mm)	Major span (mm)	Specimen length (mm)
$\leqslant$ 6.7	135	200
6.8 < 10.0	200	250
10.1 < 12.5	250	300

This flexural test is the main yardstick for determining the physical properties of grc. Reliance on this specific method of determination of these properties is no doubt attributed to the relative ease of performing the test and the reliability of the result. It expresses, therefore, the true quality of the test board and hence the production.

The variations that are monitored in the composite can be related to the LOP and MOR values as follows.

LOP (Limit of Proportionality)

The LOP strongly depends on the matrix properties and is therefore influenced by:

1. Type of matrix being used
2. Degree of cure
3. Presence of surface cracks (for example, mechanical damage and shrinkage cracking)
4. Quality of the matrix (for example, porosity and compaction) — hence the LOP obtained in quality recording tests is in the main a function of the quality and degree of cure.

MOR (Modulus of Rupture)

The MOR of a grc composite is a function of the fibre/matrix interplay and depends on:

1. Glass content
2. Orientation of the glassfibres
3. Extent of the cure
4. Composition of the matrix
5. Quality of the matrix (for example, porosity and compaction).

Glassfibre content is directly related to MOR values because the fibres increase the stress level at which the first cracks propagate and, at a later stage, carry the load across the finely cracked regions. Fibre orientation is important since there must be sufficient fibres aligned across the cracks to carry the load. The level of cure has an effect on the bond of the fibres to the matrix so that the fibres are gripped by the cement and their strength utilized.

11.5.4 Impact strength

The test that has evolved for studying the impact properties of grc is the Charpy impact test. It was originally developed by the Building Research Establishment. This method involves the measurement of the starting and finishing height of a pendulum that is caused to impact upon the sample. The difference in pendulum height yields the potential energy exhibited by the pendulum in breaking that sample. Normally, impact tests are not used in standard quality recording tests but a sketch of a suitable rig is shown in Figure 11.12.

A simple method of impact test can be performed by dropping a steel ball onto the centre of a 300 mm

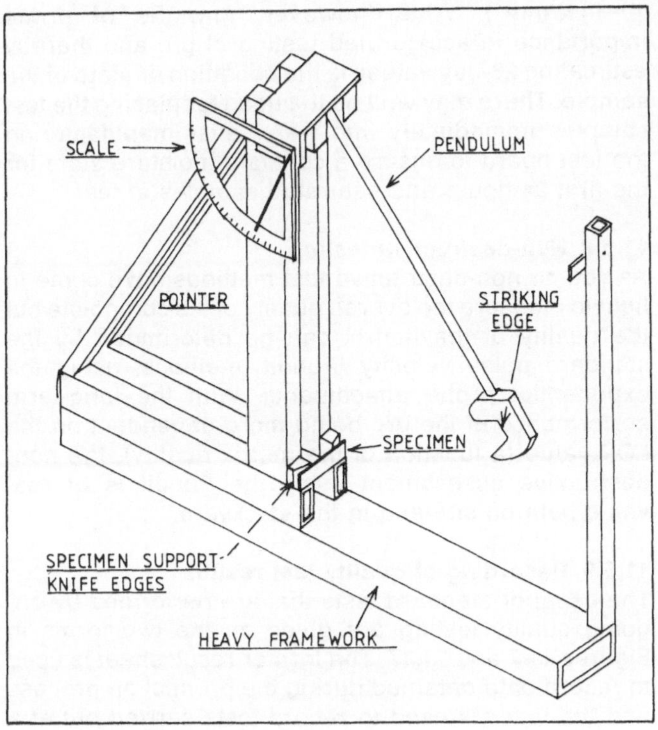

Figure 11.12 Rig for impact testing.

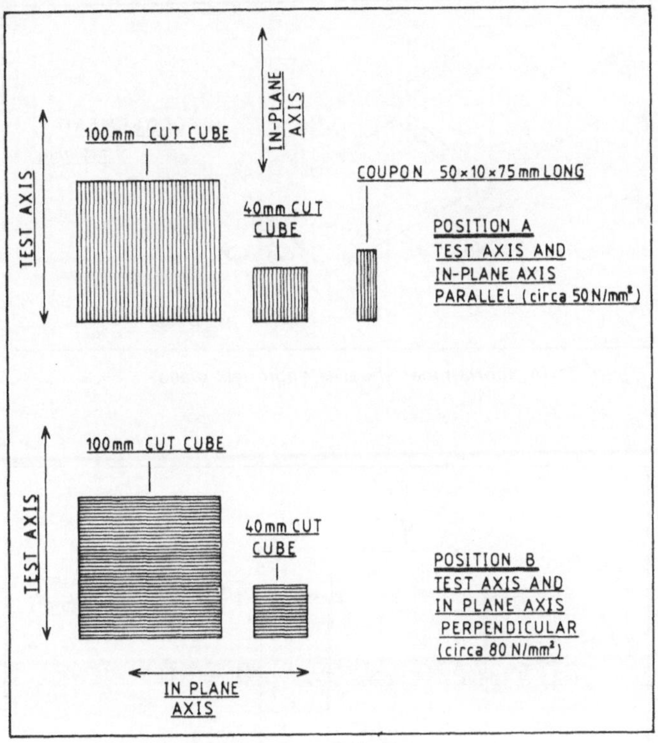

Figure 11.13 Compressive strength testing.

square sample supported on an annular ring. Either the mass of the ball is varied or the height of fall until failure occurs. The energy exhibited is equal to the potential energy of the steel ball. This test can easily differentiate brittle materials from those with a more progressive failure.

11.5.5 Compressive strength

Because grc is mainly manufactured in thin two-dimensional sheet-like components, its compressive strength is not usually measured. However, tests performed on cubes cut from material manufactured in the same manner as production components but to thicknesses 40 mm to 100 mm have given consistent data on the compressive strength. The current approach is to test standard coupons which are cut in half, viz. 75 mm × 50 mm × 10 mm thick and tested on edge.

Obviously the cutting of the coupon is critical, and grinding of the cut faces to produce parallel bearing surfaces is considered necessary. Results obtained from cut coupons yield similar values to the cut cube samples mentioned previously. Referring to Figure 11.13 for samples tested with the in-plane axis parallel to the loading axis, at position *A* they have yielded values circa 50 N/mm^2 whereas samples (only cut cubes in this case) with the in-plane axis perpendicular to the loading axis, position *B*, have yielded values circa 80 N/mm^2. Compressive strength values are a function of the matrix formulation, curing history and orientation of the fibres relative to the test axis.

11.5.6 Shear strength

Due to the anisotropic nature of spray-up grc there are a number of shear strengths. These are interlaminar shear strength, punch-through shear strength and in-plane shear strength. The following notes are suggestions for measuring these values.

Interlaminar shear strength

The test sample dimensions are shown in Figure 11.14. Cross-head speed is 5 mm/min. From the sample dimensions and the load at failure the following defines the shear strength:

$$\text{I.S.S.} = \frac{P}{ab}$$

where:
P = failure load (N)
a = sample width (mm)
b = slot spacing (mm)

Only samples that fail by shear are considered.

Punch-through shear strength

The sample is 75 × 25 mm and clamped as shown in Figure 11.15. The centre portion is punched out at 1 mm/min. cross-head speed. The shear strength is defined as:

$$\text{P.T.S.S.} = \frac{P}{2bc}$$

where:
P = failure load (N)
b = sample width (mm)
c = sample thickness (mm)

In-plane shear strength

This is a difficult test to perform but if failure occurs between the ends of the saw cuts the shear strength is defined as:

$$\text{I.P.S.S.} = \frac{P}{ac}$$

where:
P = failure load (N)
a = distance between saw cuts (mm)
c = sample thickness (mm)

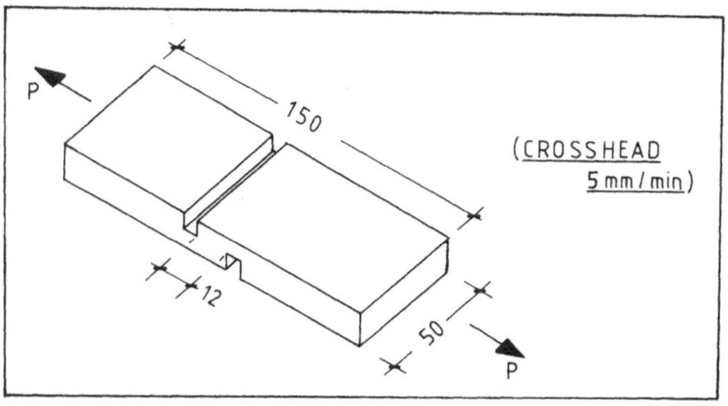

Figure 11.14 Interlaminar shear strength test piece.

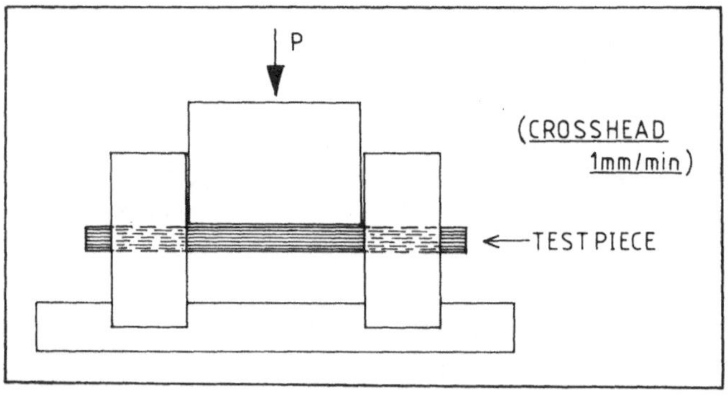

Figure 11.15 Punch-through shear strength test piece and test rig.

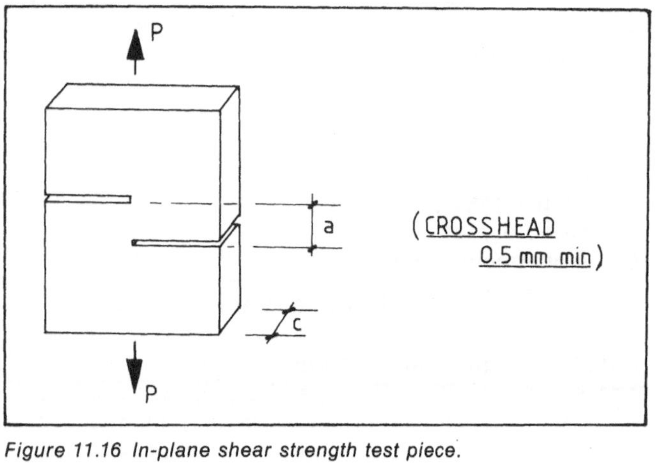

Figure 11.16 In-plane shear strength test piece.

Cross-head traverse rate is 0.5 mm/min. The test piece is shown in Figure 11.16.

11.5.7 Accelerated testing

Within recent years, accelerated testing of concrete has become a reliable means of estimating 28-day strengths at 1 day from casting. The process is carried out by accelerating the curing process (in a hot water bath) to attain a higher maturity within the test sample. The strength at 1 day after accelerated curing is then related to the 28-day strength after normal curing.

Accelerated testing could well have a significant place in grc quality recording. However, the need for accelerated curing may not be required as uniform controlled curing may be all that is necessary. This is because experiments have found that after 24 hours, the normal MOR test is 70% of the 28-day value, both these values being taken after normal curing (about 10–14 N/mm^2). What, however, may be of prime importance in accelerated testing of grc and thereby estimating 28-day values is the condition or state of the sample. There may well be the need for placing the test samples immediately into water after manufacturing the test board to ensure a constant moisture state for the first 24 hours and saturated samples at test.

11.5.8 Non-destructive testing

As yet, no non-destructive test methods have come to light to measure the overall quality of the composite but the quality of the matrix can be determined by the standard pulse velocity 'Pundit' methods, using the exponential probe attachments. With the long-term performance of the grc being more dependent on the LOP value (a function of the matrix quality), the non-destructive assessment using the Pundit is of real value both on site and in the stockyard.

11.5.9 Recording of quality test results

The usual or standard tests that are performed by on-going quality testing are given on the two forms in Figures 11.7 and 11.17. The former result sheet is used to record data obtained during the production process and the latter is used to record tests carried out at a later date. In fact two sheets of the type shown in Figure 11.17 are used, one for 7-day test results and a second for 28-day results.

A visual guide in the form of a results graph is of immense value in that it shows the trends and running standards obtained throughout a period of time, perhaps a contract run or a pre-determined number of days. The ideal graph is one that contains all the

Figure 11.17 Quality tests results sheet.

G.R.C. DATA SHEET (Composite Properties)

Date Cast: Age at Test: Date at Test:
Contract: Ref No.

Flexural Test No.	Width (mm)	Thickness (mm)	Load at first crack(N)	Load at failure (N)	L.O.P. (N/mm^2)	M.O.R (N/mm^2)	Face in Tension (S or R)
1.							
2.							
3.							
4.							
5.							
6.							

Fibre Content B/B Check ______ %
Washout Test % 1. 2. 3. 4. 5. 6.
Standard Deviation ______ N/mm^2
Coefficient of Variation ______ %

Density 1. 2. 3. 4. 5. 6.
Density (kg/m^3)

Results	L.O.P	M.O.R		L.O.P.	M.O.R.
Av.					
Max.			Std. Deviation N/mm^2		
Min.			Coeffit. of Var. %		

Compliance with Specification
Std. Deviation of last 40 Results ______ N/mm^2
Characteristic Strength Achieved ______ N/mm^2
Characteristic Strength Required ______ N/mm^2

Characteristic LOP Required
Characteristic LOP Achieved

COMMENTS

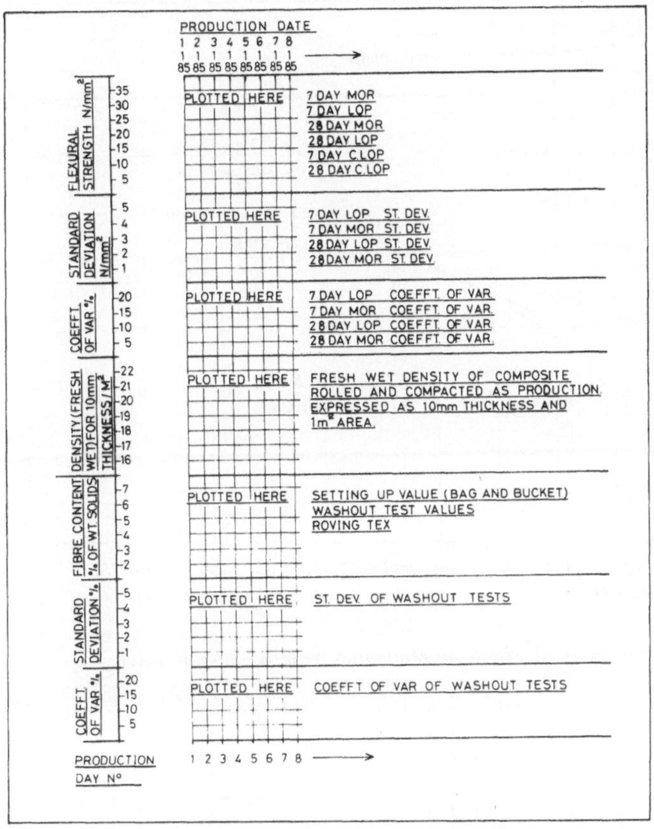

Figure 11.18 Graphical record of 7- and 28-day test results.

information obtained from the standard tests and recorded on the sheets in Figures 11.7 and 11.17. Such a graph is shown in Figure 11.18. At first it may well appear as a confused collection of data. However it not only records the results but also shows the running fibre level obtained from a wash-out test, the 7-day flexural test results and both the LOP and MOR values. It indicates, therefore, the likely properties that will be obtained at 28 days. These 28-day values are the values used for compliance testing.

Although it is not normal practice, the plotting of the running characteristics LOP (C.LOP at 7 and 28 days), yields a more true picture of the composite quality and variability, than does reliance on mere average results from daily batches of test results.

11.5.10 Suitable test equipment for strength tests

The standard test procedures mentioned in the last section, which are used as a measure of the quality of the composite, include the four-point bending test. This is used to determine the flexural strength of the material. Normally the LOP and MOR values are measured, which in the case of the former requires a load/deformation or load/time plot.

Most concrete test laboratory equipment is related to testing cubes and beams that require loads of such a value as to render the equipment insensitive at the loads required for grc flexural tests. Compressive testing and punch-through shear strength tests will, however, require such high loads as to be suited to standard cube test machines (about 20 to 60 kN).

Loads for flexural testing, interlaminar shear strength and in-plane shear strength testing will rarely exceed 2.5 kN for 10 mm thick coupons of 50 mm width (the nominal standard size).

Figure 11.19 Monsanto tensometer type 'W' with automatic graph plotting facility.

Figure 11.20 (a) Nene instruments modified Hounsfield type 'W' tensometer converted to plot a force strain diagram. (b) Nene instruments type M3000 tensile tester.

(a)

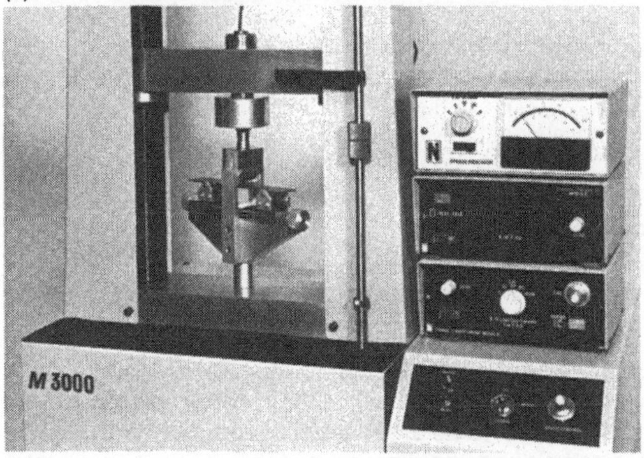

(b)

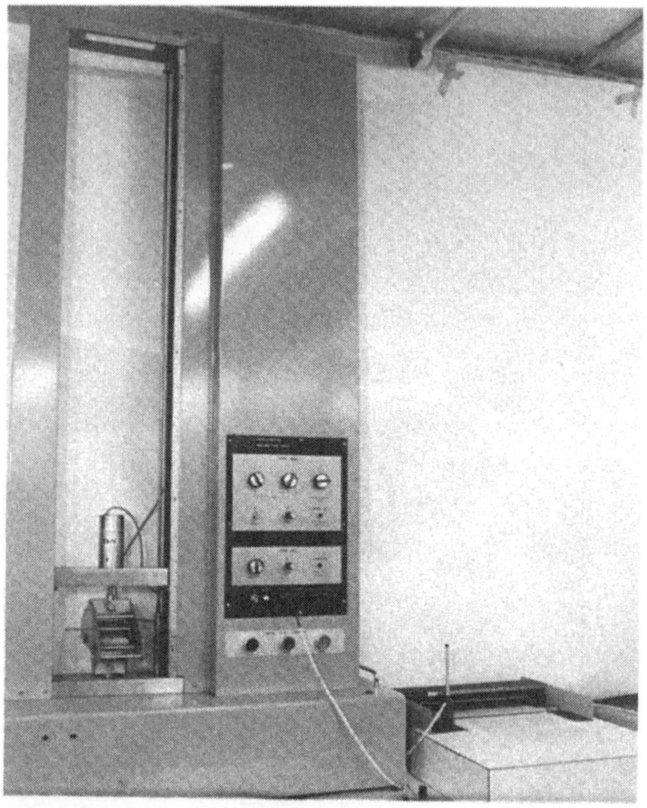

Figure 11.21 The J J Lloyd Instruments Ltd tensile testing machine.

Suitable tensile testing equipment is shown in Figures 11.19–11.21. This equipment is capable of plotting load/deformation graphs to sufficient accuracy for quality testing as shown in Figures 11.22 and 11.23. The value obtained for the load at the LOP is greatly dependent on the magnification of the deformation plot. As yet no universal test procedure has been adopted but the relationship between the values obtained on certain machines for the LOP has been noted. Possibly the two standard machines are the Monsanto Type 'W' Tensometer and the J J Lloyd T5000/SS test machine. For these machines the relationship between indicated load at the LOP is:

Monsanto (Type 'W') × 0.8 = J J Lloyd T5000/SS

This result is obtained with the instruments set as given in Table 11.3.

Table 11.3 Instrument setting.

	Monsanto Type 'W'	J J Lloyd T5000/SS
Cross-head speed mm/min.	1.58	2.00
Magnification of cross-head		
manual	8:1	—
automatic	4:1	—
Sensitivity of load cell	—	0.1
Range of chart recorder	—	1 V
Recorder chart speed mm/min.	—	300

Considerable savings in equipment costs can be achieved by adaptations to manual tensometers using the newly available DC LVDT linear displacement transducers. These are available complete with built-in electronics and require only a stable d.c. power supply

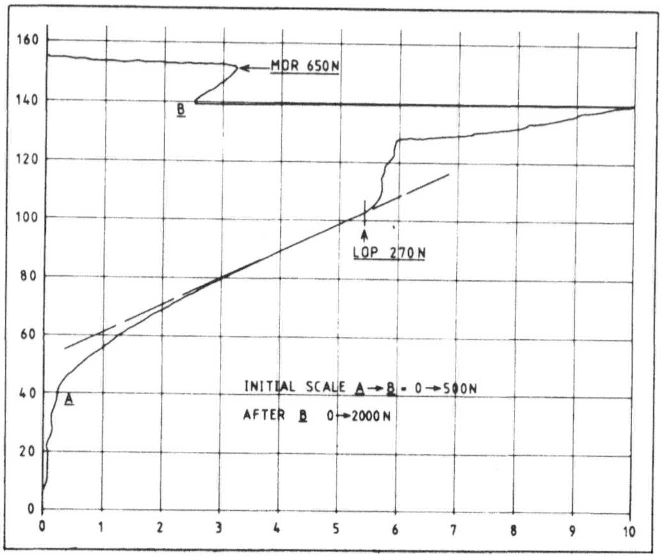

Figure 11.22 Plot of flexural test using J J Lloyd T 5000/SS equipment.

Figure 11.23 Plot from Monsanto type 'W' manual trace.

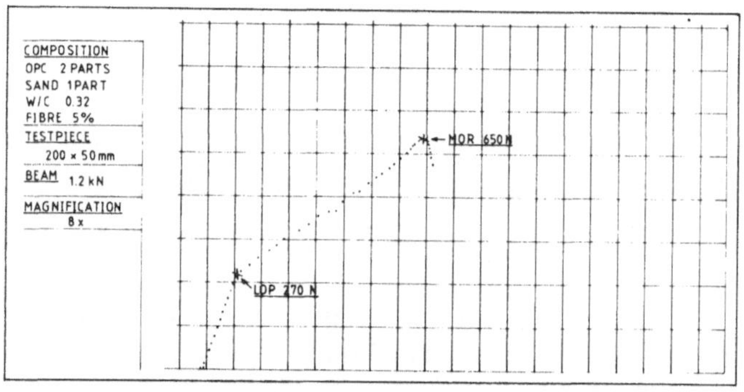

(lead accumulator) and a recorder (pen recorder). Naturally the performance of these adaptations has to be cross-checked but the financial saving makes the consideration worthwhile.

11.6 FULL-SCALE PRODUCT TESTING

In order to verify that a grc component has been designed to cope with the loads and subsequent stresses in service, and also to indicate that the quality of the manufacturing process is producing satisfactory components, a full-scale load test is carried out. This test is designed to subject the unit to the worst loading condition and to assess the performance of fixings or inter-component joints and so forth. It should determine:

1. The behaviour of the component at the maximum expected working load taking fixings and locations into consideration
2. The load deflection behaviour of the panel, noting the deviation from the initial linearity
3. The ultimate failure load of the unit and the mode of failure (compressive, tensile or shear) or the attainment of a satisfactory margin over the expected worst working load condition
4. The failure load of the fixing system or connections (in a separate test)
5. Any non-symmetrical deflections of the component
6. Creep while under test
7. The level of permanent deflection or set in the component while under test

8. The relationship between thickness of sections of the component, and the design requirements

9. The quality of the material used in the test (samples taken from compression zone if they are satisfactory)

10. The general quality of the workmanship.

11.6.1 Loading requirements and procedures

A suitable rigid steel frame is required fixed firmly to a concrete floor. The unit must be as near the floor (while under test) as possible. Deflections should be measured by dial gauges located at convenient points above the component to enable readings to be obtained safely. Deflection measurements must always be taken at the support points of the component. This allows for settlement and enables true deflection readings to be obtained. Readings should, therefore, be taken at least at the loading points and the centre point of the span of the component.

Loading is usually carried out with engineering bricks or sand bags, the unit weight of these being easily determined. Gaps between vertical columns must be left in the stacking to prevent arching under deflection, transferring the load to the ends. Alternatively, a water bag system can be used enabling the component to be loaded sideways and making deflection measurements easier and safer to perform. The water pressure is simply measured by a suitably calibrated polythene tube.

Suitable load increments, based on the design load, are applied and a pause in the loading of some 5 minutes, allows deflections to equalize and settle out. When a suitable safety margin in excess of the design load has been reached, the load is maintained for a longer period in order to register creep within the unit at the applied stress level.

The loading is then removed and permanent set is recorded. The loading is again re-applied as before and taken to either failure or a suitable level in excess of the design stress.

11.6.2 Examination of the component after testing

If the component has failed then the thickness or thicknesses of grc at the failure zone should be recorded. If failure has not occurred, then the thickness at the zone of maximum stress should be noted. The general quality of the material should be accessed visually and by cutting out coupons for LOP and MOR testing from regions that were under low stress.

11.6.3 Reporting

Load/deflection plots should be drawn for the component, indicating the LOP. The design load should also be shown on the trace.

11.6.4 Acceptance criteria

Normally the LOP is taken as 1.5 × design load but for permanent formwork this can be reduced. Ultimate failure loads should be at least in excess of 3 × design load but again permanent formwork may be lower. The ratio, failure to design loading, is taken as a function of the use and the quality of the material being manufactured.

11.6.5 Routine proof testing

When large volume production is considered, a regular load test may be advisable on units taken at intervals from the production line. For this type of test an automated hydraulic loading system can be set up with displacement transducers reading at increments as the load is applied. This test would also serve as an indicator of suspect production units or as a routine test on components that may be destined for more exacting functions where failure of one unit would lead to a high financial penalty in lost time, for example tunnel linings and mining applications.

11.7 STATISTICAL ANALYSIS OF QUALITY RECORD DATA

By taking the data obtained from quality record tests in groups of 100 consecutive days it is possible to analyse and obtain information on the interdependence of fibre content and strength levels, as well as establishing the variability in physical properties of the material. Since fibre content has a considerable effect on material costs, it is naturally worth while establishing the cost/effectiveness of the fibre addition.

It is assumed that the basic statistical analysis terms are familiar to the reader. From an analysis of hand spray-up work using a 2:1 cement/sand formulation with water/cement ratios of about 0.32 to 0.35 the following properties can be considered as average results measured by a Monsanto Type 'W' tensometer.

$\bar{x}$ MOR	23.5 N/mm^2
S MOR	3.5 N/mm^2
CoV MOR	14.9%
Characteristic MOR	17.75 N/mm^2
$\bar{x}$ LOP	10.5 N/mm^2
S LOP	2.2 N/mm^2
CoV LOP	20.9%
Characteristic LOP	7.0 N/mm^2

Determination of the correlation coefficient of the MOR against LOP indicates that a weak relationship exists <0.5. Again, analysis of the MOR against characteristic LOP yields a weaker relationship around 0.3. It is not perhaps unusual to find that the percentage fibre content is not at all related to the characteristic LOP <0.2 correlation coefficient. However, as part of a costing exercise the relationship of material cost, physical properties and fibre content produces an interesting comparison.

Analysis of 100 results taken on consecutive days was performed using hand sprayed and non-dewatered material. The process was carried out by grouping the results into blocks dependent on fibre content namely 3.8 to 4.1%; 4.1 to 4.4% and so on. Fibre content was measured by washing-out testing, and the strength tests measured after 28 days' curing, 27 of which were in water at 20°C. Figure 11.24 summarizes the results of these tests and shows that the fibre content has little or no effect on the LOP but increases the MOR with increasing fibre content. When costs are added to these data the results in Table 11.4 are obtained as well as Figure 11.25 which shows the relationship of cost against fibre content.

The indication is that little gain in design strength values is obtained by increasing fibre presence, indeed the ultimate strength and therefore the safety margin is reduced. However, the variability of the material at any particular fibre level indicates that cutting the fibre content could lead to concern in the long term. If a short life is planned it may well be more cost effective if controlled amounts of fibre at reduced levels of addition are employed.

The manufacturer must analyse his own production to enable confident, logical, cost effective and safe decisions to be made.

Figure 11.24 Summary of analysis of 100 days test results.

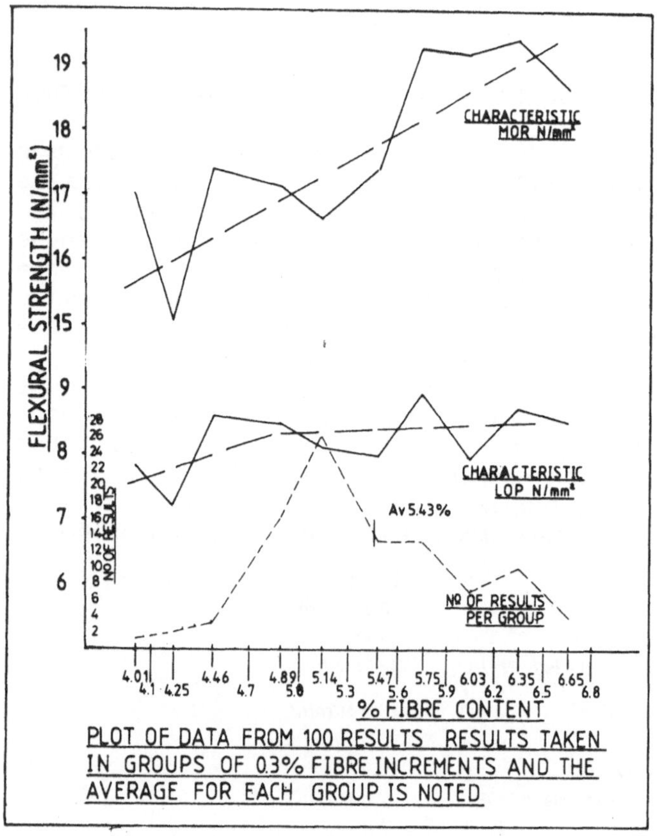

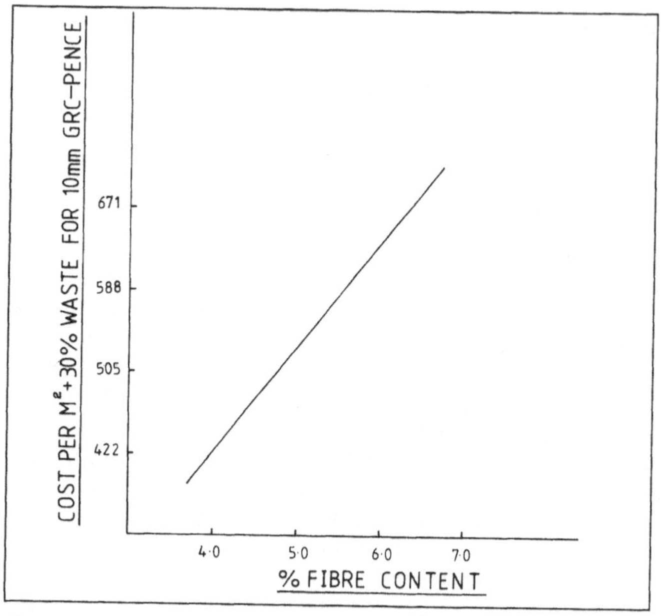

Figure 11.25 Composite cost as affected by fibre content.

Table 11.4 Relationship between costs per m^2 at 10 mm thickness and physical properties.

% Fibre content	Basic material cost including waste at 30% (per m^2 at 10 mm)	Physical properties (characteristic values)
4.0	422p	7.0 N/mm^2 C.LOP 15.6 N/mm^2 C.MOR
4.9	497p	8.3 N/mm^2 C.LOP 17.0 N/mm^2 C.MOR
6.3	613p	8.5 N/mm^2 C.LOP 19.0 N/mm^2 C.MOR

This indicates a cost increase of £1.91/m^2 at 10 mm (or a fibre content increase of 2.3% (4.0→6.3) but only increasing the design strength level from 7.0→8.5 N/mm^2 (C.LOP)).

12

Material costing procedures

Estimating as accurately as possible the quantities and therefore the cost of raw materials necessary for producing a particular product is clearly a most important aspect of planning. The following sections provide an effective guide for this purpose.

12.1 MATERIAL REQUIREMENTS

The material requirements for a product are a function of the following parameters:

1. Actual theoretical volume of the product to be sprayed. This can be determined from the drawings.
2. Extra thickness required to ensure that the minimum design thickness is achieved in production. This is naturally dependent on the following:
(a) Consistency of the production process, whether it is hand-spray or automated spray
(b) If hand-spray is used, the operator's skill which is of paramount importance in ensuring uniformity of the lay-up thickness
(c) Lowest thickness that can be achieved over the design thickness that ensures that the latter, as a minimum, is obtained with the product.
3. Product overspray. This is a quantity that is directly dependent on the perimeter of the mould and the perimeter of any block-outs in the mould to form holes, windows or the like.
4. Waste materials caused during the batching, mixing and washing-down of the plant.

12.2 GUIDE ESTIMATES OF THE PROCEEDING QUANTITIES

The theoretical volume of the product is a simple and straightforward calculation. There are, however, practical variables which have to be taken into consideration.

In order to allow for the extra thickness required to achieve the minimum design in the product, some knowledge has to be gained from the shop floor to see just how consistent the plant and the operators are in their performance. Clearly an automated spray facility in the form of a traversing head should be more consistent than a hand-spray facility. The complexity of the product will also drastically affect the quantity of extra thickness laid-up on the mould. The only possible way of estimating this value is from a background knowledge obtained by taking thickness checks on a grid pattern over products manufactured in the past and noting the variations produced. These results have to be averaged and an overall extra thickness obtained. This must be related to both the production equipment used and the operator. Also the complexity of the product produced should be noted. It is an acceptable guide to consider that 2 mm extra thickness is a minimum value. Consequently, when thin products are being contemplated of say 10 mm thickness, the extra 2 mm becomes a considerable percentage increase in material requirements over the theoretical volume of the product taken from the drawings.

The product overspray again has to be watched before an accurate value can be allowed for it in estimating. A guide for the width of the overspray in hand-spray work would be 50 mm and the thickness is probably equal to that obtained at the edge of the product. However, when overspray is folded back into the product (see *Chapter 4*) this will not produce a wastage factor that requires extra material allowance in the estimate. Because overspray remaining on the edge of the mould does not receive any roller compaction it appears to be thicker than is the case. It must, however, be considered as an equivalent fully compacted thickness.

Waste material can only be allowed for by noting the quantities during routine visits to the production and batching areas at the end of the shift. This again must be related to the day's production output.

12.3 YIELD CHECKS

Before any estimate can be obtained for the material requirement, the yield of the mix design must be known. This is of particular importance when lightweight formations are considered, or when any new filler is contemplated for use in the mix design. To obtain the yield, a value for the fresh wet density (f.w.d.) of the mix is necessary. This can be achieved by filling a suitable mould (shown in Figure 12.1) using spraying and roller compaction techniques which would be employed in production. The weight is then found for the particular volume produced in the mould. Care must be exercised when thin checks are used of, for example, 10 mm thickness. At this thickness a variation of 1 mm will have an average variation of 10% to the true f.w.d. value. It may be beneficial to use a greater thickness of say 40 mm in order to obtain the f.w.d. value and hence the yield of the mix. The surface

Figure 12.1 A mould suitable for determination of the fresh wet density of a grc spray-up mix

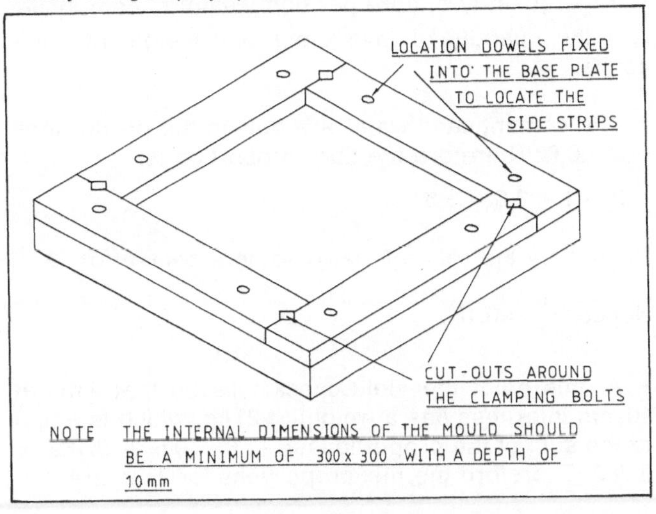

of the filled mould should, however, be flat and this is achieved by sliding a glass plate over the area.

$$\text{f.w.d.} = \frac{\text{Composite weight (kg)}}{\text{Composite volume (m}^3)}$$

Hence the yield will establish the quantity of material required to produce a particular volume of product.

12.4 WORKED EXAMPLE

The following is an example of how to set about obtaining an estimate of the materials required to produce a product as shown in Figure 12.2.

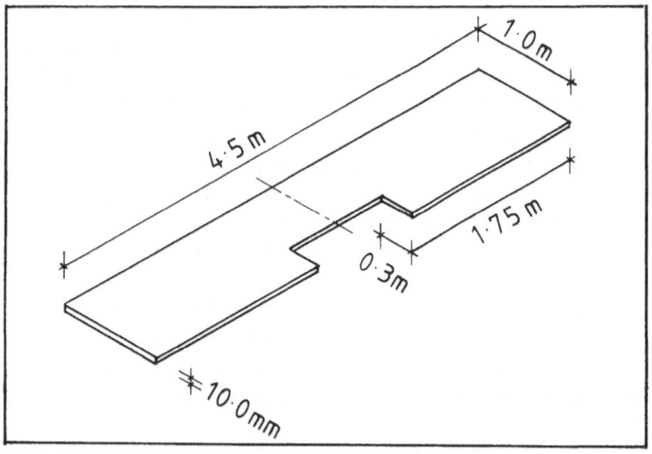

Figure 12.2 A simple flat sheet product as an example for material requirements.

• First the mix design has to be established. For this example we shall consider the following mix design:
A 2:1 mix with 5% glassfibre and a water/cement ratio of 0.3.

• Either a knowledge of the yield is available or its value has to be determined as previously discussed. We shall assume a f.w.d. of 21 kg/m^3 at 10 mm thickness (the product thickness).

• It follows by definition that a 5% fibre content is:

$$\frac{5}{100} = \frac{F}{F + M}$$

Hence $M = 19F$

where:
F = the quantity of fibre per unit weight of composite
M = the quantity of matrix per unit weight of composite.

• The cement:sand:water are now in the proportions of 2:1:0.6. Therefore the fibre proportion is:

$$\sum 2 + 1 + 0.6 = 3.6$$

Divided by the ratio of matrix to fibre content of 19:

$$\text{Hence } \frac{3.6}{19} = 0.189$$

• At this point the yield check tells us that 1 m^2 at 10 mm thickness has a weight of 21 kg, which is equal to the sum of the proportions 2 + 1 + 0.6 + 0.189 = 3.789. Therefore the mix proportions for 1 m^2 are:

Cement	:	Sand	:	Water	:	Fibre
$\left(\frac{2}{3.789}\right)21$	:	$\left(\frac{1}{3.789}\right)21$	:	$\left(\frac{0.6}{3.789}\right)21$	:	$\left(\frac{0.189}{3.789}\right)21$
11.08 kg	:	5.54 kg	:	3.33 kg	:	1.05 kg

Hence for a product as shown in Figure 12.2 of 4.2 m^2:

	1 m^2	4.2 m^2
Cement	11.08 kg	× 4.2 = 46.54 kg
Sand	5.54 kg	× 4.2 = 23.27 kg
Water	3.33 kg	× 4.2 = 13.98 kg
Fibre	1.05 kg	× 4.2 = 4.41 kg
	Total	88.2 kg

• To achieve 10 mm design thickness the spray-up is set to a target mean of 12 mm. This is an increase to $\frac{12}{10} \times 88.2\text{ kg} = 105.84\text{ kg}$.

• The perimeter of the mould in Figure 12.2 is 11.6 m and the expected overspray is 10 mm thickness by 50 mm width. This equals:

(11.6 × 0.01 × 0.05 × 2100) = 12.18 kg

Hence total material requirement is:

105.84 + 12.18 = 118.02 kg

• The total material requirement is proportioned into the material components as follows:

Cement	$\left(\frac{2}{3.789} \times 118.02\right)$	= 62.3 kg
Sand	$\left(\frac{1}{3.789} \times 118.02\right)$	= 31.15 kg
Water	$\left(\frac{0.6}{3.789} \times 118.02\right)$	= 18.69 kg
Fibre	$\left(\frac{0.189}{3.789} \times 118.02\right)$	= 5.89 kg

• As can be discovered from wash-out tests, however, the fibre content has to be set higher than 5% in order to ensure obtaining a 5% fibre content in the composite. This may require setting at 5.5%. The final material proportioning is, therefore, given as follows:

Cement	62.3 kg	[62 kg]
Sand	31.15 kg	[31 kg]
Water	18.69 kg	[18.5 kg]
Fibre	6.48 kg	[6.5 kg]

The figures shown in brackets are the final values rounded-up. They are given as those calculated most suitable for factory batching.

12.5 FINAL COSTING

Once this procedure has been undertaken in full, a cost can be put against each quantity and a total material cost established for the product. This should be checked by investigating prototypes as discussed in *Chapter 7*. No reference is made here to mould costs, man hours in production, profits, overheads and so forth as these are a function of each company's particular set-up.

13

Product range

An appreciation of the potential uses for grc is growing. With a fuller understanding of the particular benefits offered by this material, applications are becoming more varied and more widespread. In this Chapter the product range is explored under the specific subjects of cladding, composite cladding and formwork. A final section deals with significant general uses.

13.1 CLADDING APPLICATIONS

Grc cladding is particularly noted for its relative lightness and 'freedom of design shape' in comparison with conventional concrete cladding. The following selected applications illustrate, in particular, the form and size of panel that can be produced.

13.1.1 Kingston Fairfield West office complex (Figure 13.1)

Approximately 800 grc cladding panels were supplied for the 12-storey Fairfield West office complex in Kingston-upon-Thames, Surrey. The panels (Figure 13.2) were purpose made to suit the architect's design and have a fire resistance stability of 120 min., an integrity of 60 min. and insulation of 120 min.

Clients: Kingston Borough Council
Architect: Owen Luder Partnership of London

Specification	**Main window panel**
Height of panels	2780 mm
Width of panels	2180 mm
Thickness of panels	380 mm top
Thickness of panels	750 mm bottom
Weight of panels	1016 kg
Fire resistance	
Stability	120 min.
Integrity	60 min.
Insulation	120 min.
Fixing method	Non-ferrous Cast in sockets Angle brackets Adjustable channels
External finish	Smooth white (mullions: mural-plast)
Internal finish	Fairfaced
Infill	Lightweight styropore Polystyrene bead concrete
Quantity of panels	
Window panels	516
Mullions	192
Construction	Sandwich construction
Mullions	2 × 10 mm grc skins with 110 mm styropore lightweight infill
Windows	2 × 11 mm grc skins varying 380 mm at top of panel, 750 mm at base of panel

13.1.2 Stourbridge central redevelopment area car park (Figure 13.3)

Each of the 487 panels on this Stourbridge central area redevelopment car park weighs only 762 kg. The sandwich panels (Figure 13.4) are composed of 10 mm grc with lightweight infill. The components are of several different types including ribbed and flat panels.

Figure 13.1 Kingston Fairfield West office complex.

Figure 13.2 Panel detail.

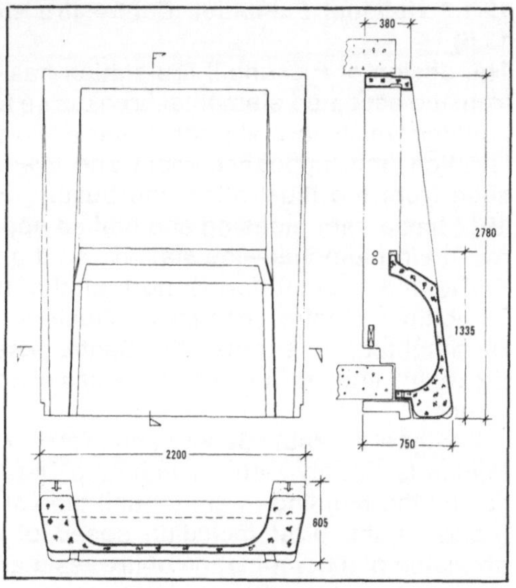

Figure 13.3 Stourbridge central redevelopment area car park.

Figure 13.4 Panel detail.

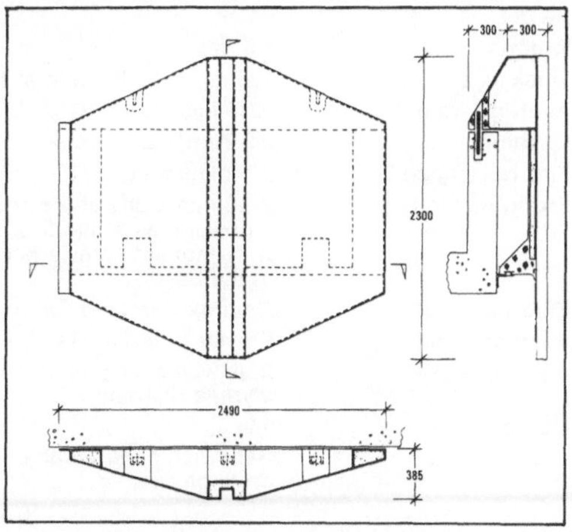

***Clients:* Dudley Metropolitan Borough in association with Nealehouse Investments Ltd**

***Architect:* J Seymour Harris Partnership, Birmingham**

***General contractor:* David Charles Construction**

Specification	**Sandwich panels**
Height of panels	2300 mm
Width of panels	2490 mm
Skin thickness	Nominal 10 mm solid grc adjacent to fixings
Weight of panels	762 kg
Fire resistance	>1 h
Fixing method	
Top fixing	20 mm dowel grouted into concrete upstand and top of panel
Bottom fixing	Adjustable channel and bonding strap
Finish external	38 mm × 12 mm ribbed and applied finish
Finish internal	Hand trowelled
Infill	Styropore
Quantity of panels	198
Construction	600 mm (top) 60 mm (base) sandwich panel comprising 2 × 10 mm grc skins with styropore infill in all sections

13.1.3 National Exhibition Centre fire station (Figure 13.5)

The design of Bickenhill fire station has moved away from the accepted elemental breakdown for this type of building which normally comprises a two storey administration unit, appliance room and tower. As can be seen from the illustration, the building completed in 1977 has a very pleasing and unified appearance with two main elements—the station and the tower.

The fire station forms part of the new National Exhibition Centre complex situated just outside Birmingham. It serves the Centre itself and also Bickenhill and the surrounding areas of Solihull District Council.

Designers, Paul Beney and Terry Hart of West Midlands County Council Planning Department, conceived the building in grc to fulfil the various requirements of the brief including speed of erection and provision of a combination of fire resistance, strength, strength/weight ratios, thermal resistance and sound reduction. Lighter structural steelwork and foundations were also achieved due to the low relative weight of grc compared with traditional materials. Panel details are shown in Figure 13.6

***Client:* West Midlands County Council—Fire Services Committee**

***Architect:* A A Wood, Dipl Arch., Dipl. T.P., FRIBA, FRTPL, FRSA, County Planner/Architect**

***Contractor:* Hampton and Sheldon Building Co. Ltd**

***Consulting Engineer:* Ove Arup and Partners**

***Quantity Surveyors:* Francis C Groves and Partners**

Specification	**First floor panels**
Height	5.2 m
Width	1.8 m
Thickness	120 mm and 160 mm at the joint
Average weight	1219 kg
U. value	0.9 W/m^2/°C
Fire resistance	2 h minimum
Fixing methods	2–16 mm studs at the roof level 2 dowel pins at the first floor 2–12 mm set screws and angle brackets at the base
External finish	Camrex-Camdura Satin
Internal finish	Plaster board sealer
Infill material	Lightweight styropore (80 mm), polystyrene (20 mm)
Quantity	144
Construction	Sandwich panels with back-up wall Forticrete blocks

Figure 13.5 National Exhibition Centre fire station.

Figure 13.6 Panel detail.

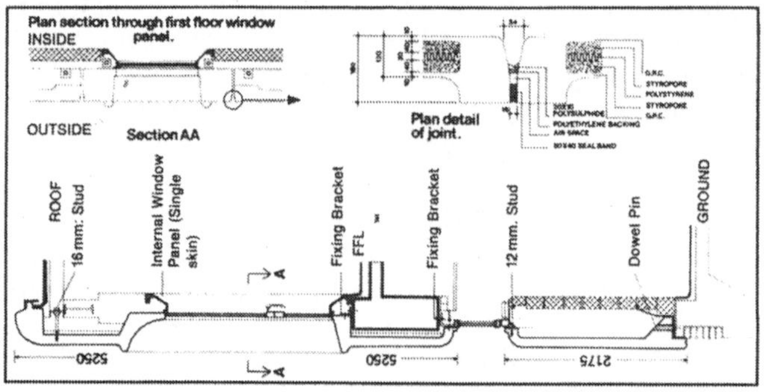

13.1.4 St Helens central area re-development phase 2 (Figure 13.7)

In the St Helens central area re-development phase 2, the architect wanted to relieve the brickwork elevation for the tall top storey. Appearance was not the only criterion—construction time and maintenance-free finish were essential. Grc provided the perfect solution. Panel detail is given in Figure 13.8.

***Client:* St Helens Council**

***Architect:* Kingham Knight Associates, St Helens**

***Main Contractor:* George Wimpey & Co. Ltd**

Specification	
Height	
Vertical	3.675 m
Slant	3.689 m
Width	992 mm
Thickness	Nominal 10 mm with stiffening ribs
Weight	182 kg
Fire resistance	Non-combustible to BS 476: Part 4: 1970
Fixing method	
Top	'U' shaped bracket. Cast in sockets and set screws
Bottom	'T' bar. Cast in sockets and set screws
External finish	Copper colour 'Minkote' (applied in situ)
Internal finish	Float
Infill	None
Quantity	66 as illustrated 28 specials of varying length 8 corner panels 102 total
Construction	10 mm single skin panels with 100 mm × 100 mm vertical rib stiffeners encapsulating fixing sockets as illustrated. All panels have ribbed profile
Lifting apparatus	2 no. 75 mm M16 lifting sockets cast in top of panels

Figure 13.7 St Helens central area re-development phase 2.

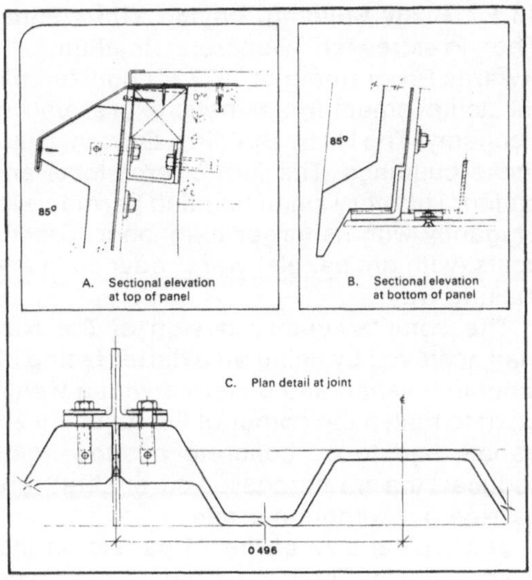

Figure 13.8 Panel detail.

13.1.5 Hotel closet modules, Roseville, Minnesota (Figure 13.9)

The Cricket Inn Economy Motel, Roseville, Minnesota is a 116-room motel designed as a prototype for future Cricket Inn motels to be erected in the midwest USA by Turnpike Properties. For this reason the design incorporates many innovations which can be used economically on future units. In particular the unique closet appendages make possible a space-saving room layout (Figure 13.10) which offers more openness than is usually found in economy class motels.

Plans originally called for precast concrete closet modules stacked one above the other creating a three-dimensional vertical spandrel comprising four storeys. The weight of the precast modules would have been 3181 kg each requiring outboard footings to support their load. The architects, however, selected an alternative solution—thin shell modules of spray-up grc. The wall thickness averaged 10 mm and the units were 1987 mm × 2725 mm × 587 mm. They weighed 272 kg and were therefore light enough to be supported by the main structure. Consequently the outboard footings originally designed for precast concrete were eliminated.

Sixty identical units were required and these were made in three plastic-faced plywood moulds. The sides of the moulds opened on hinges for easy stripping. Wood 50 mm × 100 mm forming stud assemblies for

Figure 13.9 Hotel closet modules, Roseville, Minnesota.

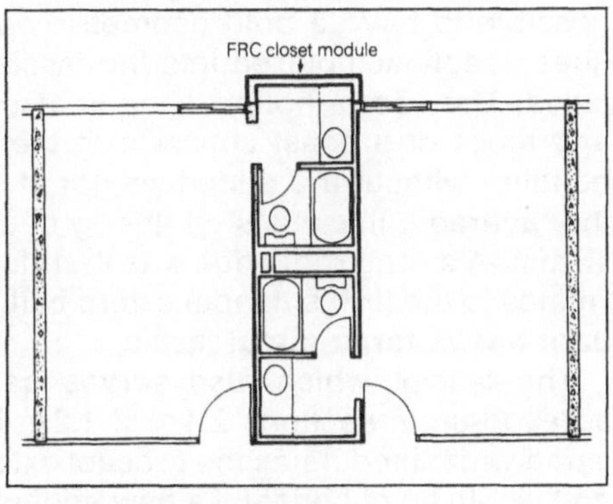

Figure 13.10 Floor plan of typical closet modules.

the walls, ceiling and floor of the closet were secured to the inside of the sprayed-up shell. This wood assembly permitted fast installation of the insulation and dry wall after the modules were erected.

Owner/Developer: ***Turnpike Properties Midwest Inc., Minneapolis, Minnesota***
Architects: ***Miller Hanson Westerbeck Bell Architect Inc., Minneapolis, Minnesota***
General Contractor: ***Gunnar I Johnson and Son, Inc., Bloomington, Minnesota***
Manufacturer and Installer of panels: ***Glas-Con, Inc., Bloomington, Minnesota***

Number of units	60
Size of units	1987 mm × 2725 mm × 587 mm

13.1.6 Fascia panels, American Guidance Services warehouse, Circle Pines, Minnesota (Figure 13.11)

An ordinary utility-looking design is acceptable for warehouses situated in non-residential areas. The new 2000 m^2 warehouse for American Guidance Services, however, was designed for a residential area of Circle Pines, Minnesota, and on a site adjoining the firm's headquarters. The two buildings are now connected by an enclosed walkway.

Project architect Ken Peterson, of Sövik Mathre Sathrum Quanbeck Architects, designed a rectangular 30 × 60 m structure with raked texture concrete panels for the exterior. Precast piers were set outside the building perimeter to rise above the roof line and provide visual relief to the two long sides.

Fascia panels of grc border the top perimeter and cap the piers. Initially, pre-finished steel panels over a wooden frame were considered but eventually discarded in favour of spray-formed grc. This made it

Figure 13.11 American Guidance warehouse, Circle Pines, Minnesota.

possible to have a bold geometric pattern of shadow lines deeply sculptured into the fascia's exposed surfaces. The use of hollow shapes also gave the depth and mass of precast concrete to the top edge of the building without the added weight of solid concrete—the average thickness of the grc units being only 12 mm. A strong roof edge was therefore achieved and relates to the firm's headquarters building which has a dark brown terne metal fascia.

The fascia, which also serves as a coping, was sprayed-up in sections 2.4 m × 1.2 m high. This is the same width module as the precast exterior wall panels and would be of benefit if a new addition was built. The wall extends as a parapet above the roof and the roof membrane wraps up and over it. The grc fascia-coping section, which is about 380 mm deep, hooks over with an overlapping back lip which protects the panel and roof edge. The fascia top slopes down to drain to the roof.

The lightweight hollow sections were easily hoisted and could just sit in place until fasteners were applied. Each section is fastened through the face at three points with stainless steel bolts. These are in recessed sockets reinforced with cast-in steel washers and concealed in the shadow line grooves. The backs are also fastened at three points by lag screws into a wood nailer at the top of the parapet. It must be noted that the fascia rests on the top of the wall panels, so the face-fasteners are not load-bearing. Details of panel fixing are shown in Figure 13.12.

Figure 13.12 Panel fixing detail.

A clever mould design enabled the six-sided pier caps to have openings only where they slip over the top of the pier and lap the fascia. This was achieved by using a single form built in plastic-faced plywood which resembled a suitcase with hinges. It was laid open for spraying and rolling the grc mix and then closed. Finally a man squeezed through the opening to place additional grc over the hinged joint.

Colour was used to provide the finishing touches. The fascia and caps were painted a medium brown. This provides a contrast to the light tan wall panels and compliments the colour scheme of the main building.

Once completed the pier caps were slipped over the pier and face fasteners in a manner similar to the fascia. In both cases, this method of attachment allows easy removal and re-use when an addition would connect an old to a new structure.

***Client:* American Guidance Services, Inc., Circle Pines, Minnesota, publisher of educational material**
***Architect:* Sövik Mathre Sathrum Quanbeck Architects, Northfield, Minnesota**
***General Contractor:* Gunnar I Johnson and Son, Inc., Bloomington, Minnesota**
***Manufacturer/Installer:* Glas-Con, Inc., Bloomington, Minnesota**

13.1.7 Nizny Building, Dayton, Ohio (Figure 13.13)

The Prestressed Concrete Institute (PCI) Annual Awards Programme in 1977 recognized nine buildings for achievement in aesthetic expression, function and economy. The Nizny Building, Dayton, Ohio, was one of these buildings. The jury of architects and engineers stated 'This very small building harmonizes simply and elegantly with its larger neighbour. Erection time and costs (with grc panels) were reduced in a very forward looking way.'

The complementary design of the Nizny Building was achieved by using an exterior facing of grc panels, smooth finished and coloured with a weather proofing stain to match the colour of the adjacent structure. The panels had to be concrete but the lightness of grc reduced the frame costs and enabled erection to be carried out without a crane.

The typical size of the 73 panels on the building is 2.4 m × 1.2 m. They are 10 to 12 mm thick and weigh 100 kg each. It is worth emphasizing that, if reinforced concrete had been used in the conventional 125 mm thickness, the panels would have weighed 1000 kg.

Grc was sprayed into plastic-coated plywood forms to create the smooth face. Anchoring inserts were set in the inside panel surfaces for bolt connections and lengths of steel angles bolted to the inserts. The angles were then anchored to the building frame by welding or bolting (Figure 13.14). Slots in attachment hardware allowed for adjustment during installation. The attachment procedure is shown in Figure 13.15.

Besides saving the use of steel, the lightweight grc panels provided installation cost savings. Four men installed all the panels in only 6 days, requiring simply a small roof-mounted hoist to lift the panels into position. After installation a foam insulation was applied to the back surfaces. The interior was finished with stud walls, glassfibre bolt insulation and gypsum board.

***Office:* Retail Building, Dayton, Ohio**
***Client:* Developer Builder: M H Nizny Co., Dayton, Ohio**
***Architect:* Gary L Herald, Dayton, Ohio**
***Manufacturer and erector of panels:* Concrete Technology, Inc., Springborough, Ohio**

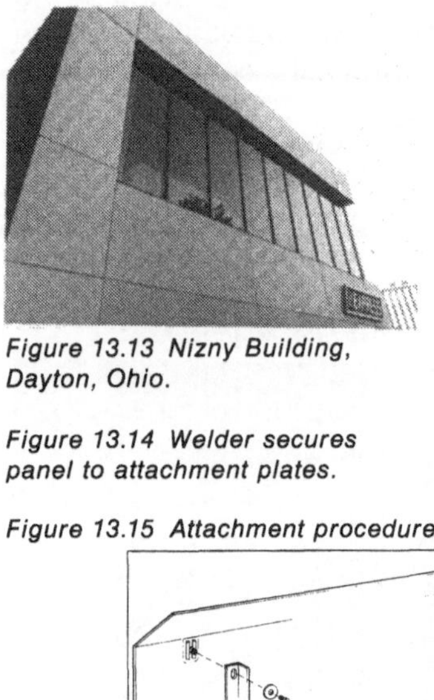
Figure 13.13 Nizny Building, Dayton, Ohio.

Figure 13.14 Welder secures panel to attachment plates.

Figure 13.15 Attachment procedure used on grc panels.
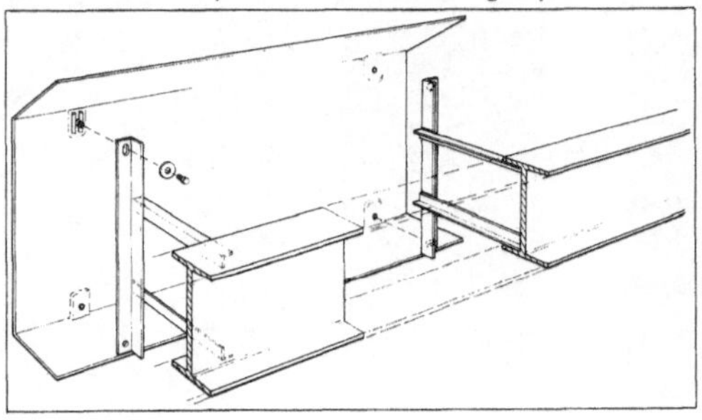

13.1.8 Kobacker Stores, Inc., headquarters and distribution centre (Figure 13.16)

Kobacker Stores, Inc., Columbus, Ohio, have built a new stores and warehouse complex. The Company wanted to create a 'positive visual presence' in the community and therefore chose a high traffic site facing the main roadway system. To achieve the aesthetics demanded a deep reveal was required that would cast heavy shadows. As a bonus the contours chosen created additional strength in the same way as the configuration of a boat gives strength to the hull. Grc has been used as cladding on this 3300 m^2 office building.

The grc aggregate-faced panels, each 2.4 m × 1.5 m, have a rectangular recess in the centre which, as required, was used for solar bronze windows. When windows were not needed the recesses have the same aggregate surface as the rest of the panel.

For increased efficiency, the firm built eight forms of plastic-faced plywood to produce the 264 exterior wall panels and four forms to produce the 121 fascia panels. Forms for the narrow flat panels that frame the windows of the lobby and cafeteria were designed to be blocked off for several panel lengths.

Figure 13.16 Kobacker Stores, Inc., headquarters and distribution centre.

Figure 13.17 Back view of solid panels showing fixing arrangements.

To produce the panels, the aggregate face was first placed in the forms. The resultant 10 mm thickness was achieved by building up layers of grc. For the large flat panels, including the fascia, rigid foam strips were added and sprayed over with additional grc to provide stiffening ribs. The grc was also built-up around the moulded-in attachment sockets.

After a 1-day initial cure, crews stripped the forms and water washed and brushed the surfaces to expose the aggregate. Seven-day wet curing followed before a high gloss acrylic sealer was applied. The sealer makes the panels easier to maintain and brings out the colour and texture of the aggregate surface.

At several stages of construction the light weight of the grc panels proved excellent in cutting costs. An estimated 5 t or more of steel framing were saved at the outset because grc panels do not require the massive structural requirements necessary for conventional precast. Figure 13.17 shows the back view of solid panels with fixing arrangements.

The overhanging fascia panels, which weigh 155 kg each and project 1.5 m, are supported by the light steel roof frame. Even the heavier 250 kg exterior wall sections were installed using only a light capacity 'cherry picker' crane.

In comparison with precast, the panels are also very much thinner and the contours give additional space for insulation. Indeed, 90 mm of glassfibre insulation was installed behind the recesses and there was still space for more. Space was also available to run electrical and telephone conduits.

***Client:** Kobacker Stores, Inc., Columbus, Ohio*
***Architect:** Samborn, Steketee, Otis and Evans, Inc., Toledo, Ohio*
***Manufacturer and Installer of panels:** Concrete Technology, Inc., Springborough, Ohio*

13.1.9 Lutheran Social Services office building, Minneapolis, Minnesota (Figures 13.18 and 13.19)

Grc panels, two storeys tall and 3 m wide, provide a neatly sculptured exterior for the Lutheran Social Services office building addition in Minneapolis. The 1700 m^2 'U'-shaped addition houses the Lutheran Social agency's state administrative offices. It also includes a 'centrum' multi-purpose conference centre seating 150, a mailing centre and a dining area which serves the entire complex.

This addition is a two-storey structure on a slab on grade. A portion of the lower floor goes below grade. The hollow precast concrete plank floor and roof deck bear on masonry end-walls and interior steel columns. The 'centrum' portion projects wedge-like above the roof line and, on the inside, exposes a wood roof deck on laminated wood beams.

Figure 13.18 Lutheran Social Services office building, Minneapolis, Minnesota.

Figure 13.19 Deep section of panels.

Walls which run at right angles to the brick bearing walls, including a courtyard inside the 'U', are faced with the grc panels. These are 6 m high for the portions where both storeys are above ground. The weight of the panels is supported by the foundation wall. The panels average 18 mm in thickness and the largest ones weigh only 900 kg. A typical precast concrete unit in this design would weigh about 6360 kg. The grc panels could therefore be handled in the plant with an ordinary fork lift and erected using a Bantam crane.

The designer established the boldly contoured shape following general criteria given by the manufacturer. The panels create a continuous line across the top, incorporating a parapet, a roof-like slope and then a vertical bond. Sections divide into vertical fins which separate shadow-box recesses for windows or spandrels—two recesses per storey. The panel is enhanced by the use of Trinity Warm Tone cement which produces an over-all tan colour.

The manufacturer used a single wood form of plastic-faced plywood to provide a variety of panel configurations. The full form gave four recesses. Each could become an integral aluminium window frame or spandrel. The form was blanked off for the one-storey panels. A simple form change produced two-storey panels, used where the lower floor goes below grade. Vertically, the form incorporated both left and right corner returns. One or both were blanked off for intermediate joint conditions.

Steel attachments were placed right into the wet material and additional grc was applied to cover. After about 16 hours of curing time, panels were stripped from the form and cured wet for 7 days. The grc panels were then ready for installation. Erection of these panels was relatively fast, an average of about one panel per hour. When erected the panels were ready for the application of furring strips and foil-backed gypsum board interior finish. A urea-formaldehyde expanding foam was pumped between the studs and into the voids of the panels, producing an extremely well-insulated exterior wall.

Client: *Lutheran Social Services of Minnesota, Minneapolis, Minnesota*
Architect: *Sövik Mathre Sathrum Quanbeck Architects, Minneapolis, Minnesota*
General Contractor: *Gunnar I Johnson and Son, Inc., Bloomington, Minnesota*
Manufacturer and installer of panels: *Glas-Con, Inc., Bloomington, Minnesota*

13.1.10 Renovation cladding, Sarasota, Florida (Figure 13.20)

The new First Professional Building in Sarasota, Florida is actually a 40-year old grocery store! It was rescued from demolition to become a contemporary office structure.

The grc panels specified to modernize the exterior of the building form three-sided cantilevered vaults which overhang the perimeter of the 780 m^2 building. Between the vaults are deep recesses. At the top of the recesses smooth-faced grc pilaster caps give the appearance of a heavy continuous support beam.

The architect designed the projecting segments to provide interesting depth and geometric shadowing to an otherwise rather flat exterior (Figure 13.21). Grc was selected as a backing to a stucco exterior, because of the deep-ribbed, rough texture it was possible to achieve (Figure 13.22). Normal stucco treatments tend to self-destruct in the Florida rain and sun.

Figure 13.20 Renovation cladding, Sarasota, Florida.

Figure 13.21 Detail of cladding components.

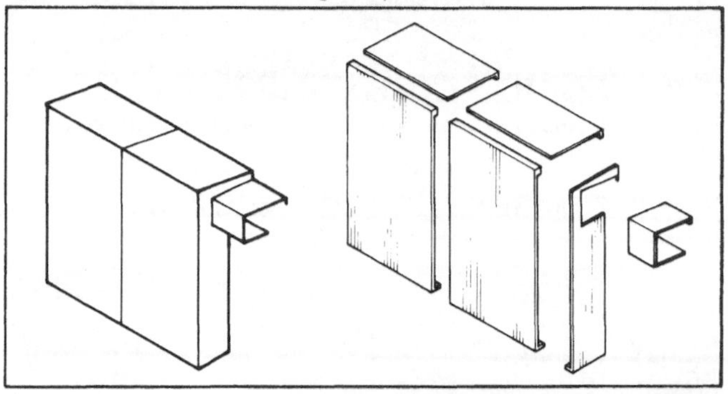

Figure 13.22 Deep ribbed texture on thin wall grc.

Grc panels were also designed to overlap the top of a 2 m high parapet which had provided a high false wall around the original building. The grc top proved completely weather resistant during severe rainstorms. If the stucco alternative had been used separate built-up roof sections would have been required over each unit.

Fitting panels to an old building such as this can be a problem. The size of each panel segment had to be adjusted to fit the varying and out-of-square measurements. Altering the forms, however, was easily done by moving and refastening wood stops and then spraying-up each part.

The vaults were assembled from three panel types—face, sides and roof. The largest units, about 2 m × 4 m were paired to make the face of each segment. Roof panels fit under lips formed in the upper edge of side and face panels, providing caulk joints. The installers built the supporting steel framing then attached panels individually. The ribbed face was unique in that an ABS plastic 'fractured fin' form linear was used in the form to create the ribbed texture. To achieve the grainy finish the sand-cement mix was poured and the mould vibrated to create a layer 3 mm above the 25 mm × 25 mm ribs. The grc mix was then spray applied to an additional thickness of 12 mm. Foam inserts were placed and sprayed over to make reinforcing ribs. Solid 62 mm mounds of grc were built up around the anchor attachments. Finally, to provide a more even colour tone, a waterproof sealer was applied to the panel surface.

***Client:* Goar and Endriss, CPA Sarasota, Florida**
Antony Sypula, Insurance/Pension Consultant, Sarasota, Florida
***Architect:* Arthur Mead. AIA, Sarasota, Florida**
***General Contractor:* John Hartenstine General Contractor, Sarasota, Florida**
***Manufacturer and erector of panels:* Cochran Concrete Company, Sarasota, Florida**

13.2 COMPOSITE CLADDING PANELS

Additional stiffness, load carrying capacity and substantial improvements in insulation can be achieved by composite panel systems. These combine grc surface skins of 6–10 mm thickness with a structural foam core that bonds the grc skins and provides a composite system that has distinct and viable properties. The panel system is shown in Figure 13.23.

The composite cladding principle has been developed by a Dutch Company Veldhoen-Isolatie B.V., Raalte, into a two-fold product system, namely:

- Fenox cold store panel
- Ranox facade panel.

Figure 13.23 A grc/poly-isocyanurate foam composite panel.

Figure 13.24 Fenox cold store building.

Figure 13.25 Cross section of the Veldhoen-Isolatie panel.
1 Mechanically fixed synthetic layer (PVC base).
2 Fenolit glassfibre reinforced concrete (grc).
3 PIR foam (poly-isocyanurate).

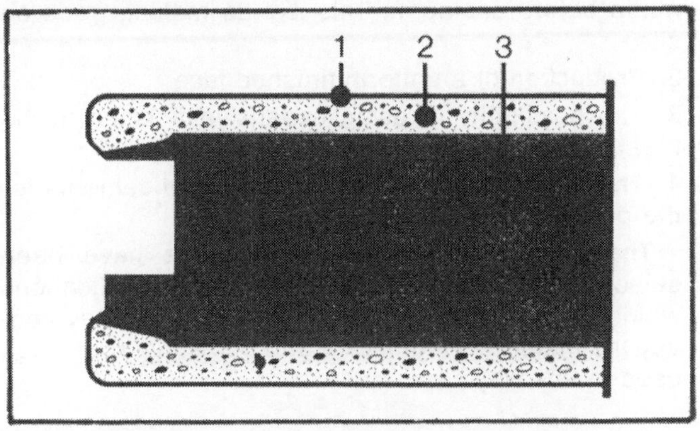

13.2.1 Fenox cold store panel (Figure 13.24)

This panel (Figure 13.25) employs a poly-isocyanurate (PIR) foam core sandwiched between 6 mm thick sheets of grc. The surfaces are coated internally and externally with a mechanically anchored synthetic layer (PVC base). The Company offers 12 standard colours on the outside elevation of the panels and 1 standard colour for the inside elevation.

The panel is manufactured from a regulated set cement that enables the highly mechanized production plant to fabricate and handle the product quickly without waiting for strength build-up, as is the case with oPc and RHPC grc. Product manufacture and demoulding is contained within a 15 min. cycle as opposed to 16 hours when using more conventional cements.

The manufacturing process is capable of producing cladding panels up to 12 m length in 0.3 m modules, in widths of 1.2 m. The thicknesses offered are 120, 160 and 200 mm. Connections between the internal steel frame and the cladding panel are made by means of a special aluminium profile. The composite has an approximate weight of 38 kg/m^2.

13.2.2 Ranox facade panels (Figure 13.26)

With the introduction of the Fenox cold store panels, other cladding applications became possible, and the Ranox facade panels were developed.

Figure 13.26 Ranox facade panels.

The major variation between these two systems is that the Ranox panels are manufactured from 10 mm skins of grc in two thicknesses 100 mm and 120 mm in widths of 1.2 m and lengths up to 12 m. The joints are finished in a special rubber profile (EPDM).

The surface of the Ranox panel can be profiled, both inside and out, producing a weight of 55 kg/m^2.

13.3 FORMWORK APPLICATIONS

Due to the inherent properties of grc it has considerable potential as a permanent rather than re-usable formwork material. In particular advantages offered are:

1. Pre-inspection and approval of the units can be made before casing-in. This avoids making good on site
2. Production of a uniform finished face
3. The possibility of using expensive finishes in the ideal conditions of a precast works
4. The employment of white or coloured cements for the grc and grey oPc for in situ work.

The following formwork applications have been selected in order to illustrate general applications which have proved particularly successful. They convey the material's significant and beneficial role when used in this way.

13.3.1 Parapet shutters

In order to achieve a clean sweeping curve line to a bridge, the designer prefers to install the parapet after the main structure has undergone its initial settlement. This means that conventional shuttering in the form of timber or steel formwork requires either the re-erection of falsework, or the retention of the original falsework system while the settlement occurs. Both alternatives are expensive. The two illustrations, Figures 13.27 and 13.28, show solutions to this problem offered by grc permanent formwork. In Figure 13.28 the forms can be seen held in place while the concrete is poured and compacted by a cantilever timber falsework system. When the in situ concrete has hardened sufficiently, the falsework can be removed and re-used further along the bridge. This produces an effective solution to the problem which is both technically satisfactory and economical.

Figure 13.27 Permanent formwork parapet units on an M4 access bridge, South Wales.

Figure 13.28 Parapet units being installed on the M4 Loughor Bridge, West Glamorgan, South Wales.

13.3.2 Bridge decking

Bridge decking formwork has proved to be a success story in its own right. From the early applications of high alumina cement and 'E' glassfibre using single skin profiled sheets through to the sandwich panel systems in vogue today, the product has evolved to fulfil the needs of the bridge designer in a number of ways.

Figure 13.29(i) shows a bridge deck under construction using Type 1 formwork as detailed in Figure 13.29(ii) and applied in Figure 29(iii). This system together with Type 2 formwork is used almost universally to span between bridge beams, up to 1.5 m O/A span. Type 1 formwork is produced as a single skin, single-spray pass panel, using specially manufactured timber moulds, formed to provide the trapezoidal hollows that strengthen the panels. Type 2 formwork is produced as a two-pass system but uses relatively simple flat sheet moulds and forms the trapezoidal hollow by encasing polystyrene formers that are laid down onto the first layer of grc. The second pass encases these and bonds them in place.

Greater spans can be accommodated by the Type 3 three-pass system. In this case further polystyrene formers are laid down in between the ones described for Type 2 formwork and a final layer is sprayed over the top. This system is expensive, however, due to the production time and involvement caused by the two separate profiles of polystyrene that are required and the three passes of grc.

Finally, Type 4 has been to-date a development exercise that has set out to achieve the load carrying capacity of Type 3 formwork, at a low production cost. As yet, however, it has not been used in a commercial application.

Ribbed grc decking has the advantage for the bridge contractor. The tops of the hollow ribs can be used to support the distribution steel and, with wire slings from the distribution steel, offer support for the main reinforcement bars which can be located in the bottom of the hollows. The hollowed profile reduces the bridge deck weight and gives a satisfactory profile to the underside of the bridge.

With due consideration of the highly impermeable nature and durability of grc, the Norwegian Bridge Committee of the Norwegian Concrete Association has with foresight recommended that if grc is used as bridge decking the cover to main reinforcement need only be 10 mm when employing 6 mm grc panels. The distribution steel in this case is considered acceptable when resting on the tops of the formed hollow ribs. In order to understand this ready acceptance Figures 13.30 to 13.32 should be studied. These show quite clearly the benefits and merits of grc as a surface skin to structural concrete.

Figure 13.29(i) Grc bridge decking being laid down to act as permanent shuttering on a road bridge in Norway.

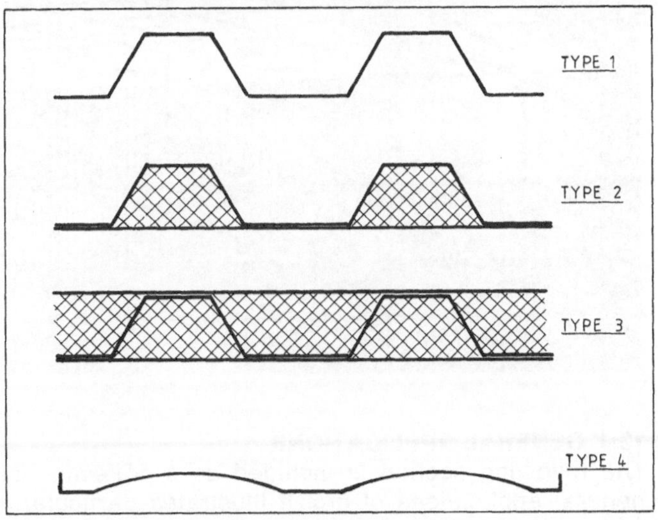

Figure 13.29(ii) Types of bridge deck formwork.

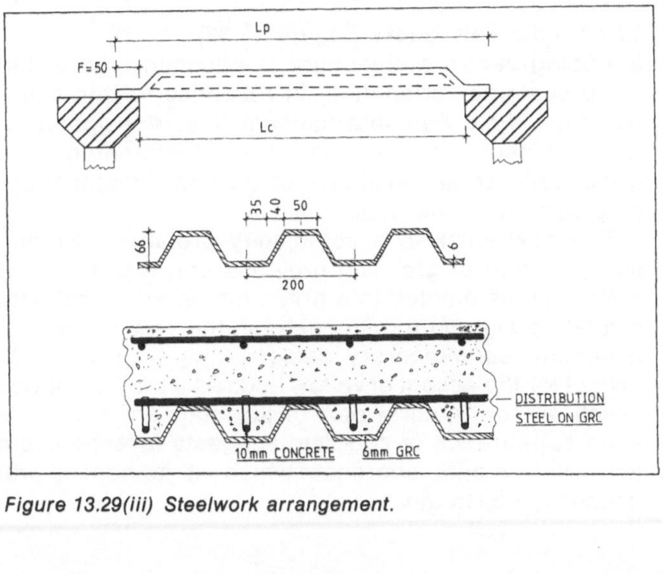

Figure 13.29(iii) Steelwork arrangement.

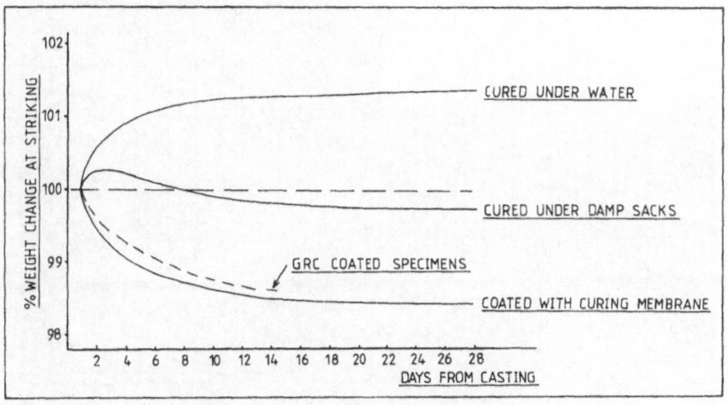

Figure 13.30 Illustration of grc as curing membrane.

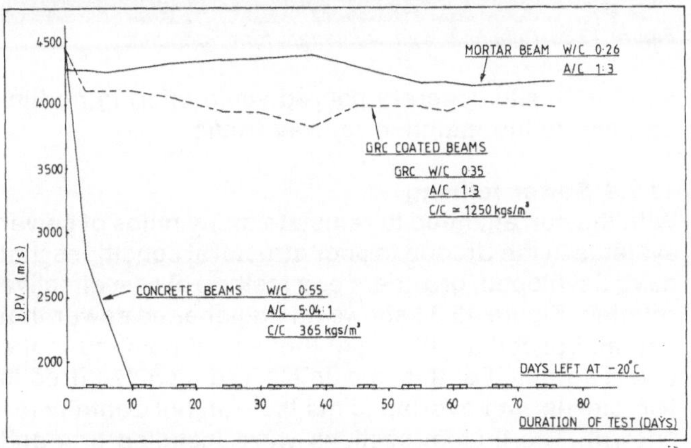

Figure 13.31 Plot of ultrasonic pulse velocity with freeze/thaw (−20°C +20°C).

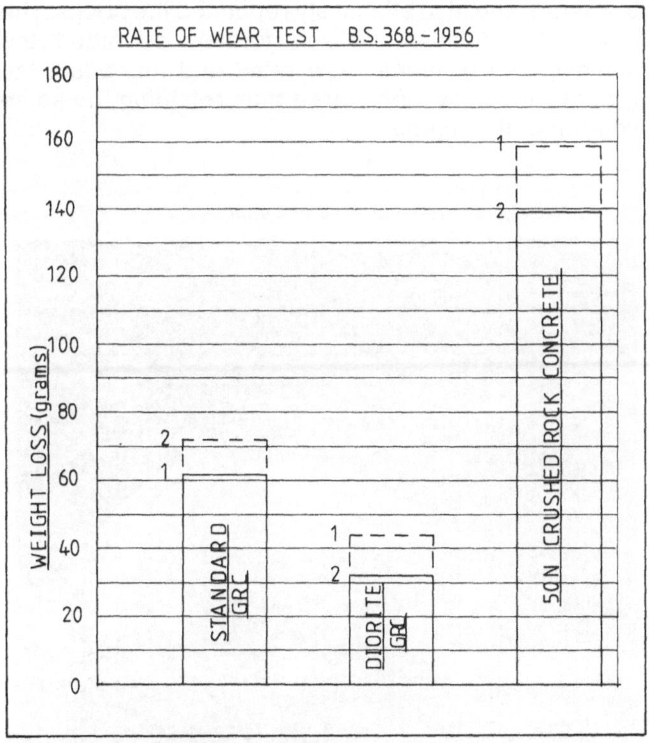

Figure 13.32 Abrasion resistance test.

13.3.3 Retaining walls

Some design solutions using grc permanent formwork shutters have been developed. Figure 13.33 shows one application that uses a concave shutter panel with substantial polystyrene rib formers encased into the back face. This system enables the shutters to be erected

Figure 13.33 Retaining wall permanent formwork units.

and the in situ concrete poured while offering no hinderance to the mainline railway traffic.

13.3.4 Sewer relining

With the current need to reinstate many miles of sewer systems in the UK due to poor structural conditions that have developed, grc offers one method of cost effective relining. Figure 13.34 shows an egg-shaped sewer that had all but collapsed due to the lack of pointing in the invert portion. For this, grc units were manufactured in two pieces with overlap joints that run horizontally (as shown). The bottom sections were installed first and then the tops positioned into the bottoms. Following this the void behind the lining and the brickwork was injected with cement grout which bonded the grc to the brickwork and also effectively repointed the bricks. The relatively smoother surface formed permitted the achievement of higher flow rates and this offset the reduced cross-sectional area thus retaining the same volume of throughput.

Figure 13.34 Sewer relining—two piece units.

13.3.5 Permanent coffer units

The principle of a coffered ceiling is well understood, but conventional methods employ grp shutters that have to be struck and subsequent defects made good. The units also come only in standard sizes offering little aesthetic scope for interior design considerations. The illustration in Figure 13.35 shows a grc coffered ceiling that was formed from units of approximately 1.2 m cube. The tops of the units housed lighting and service ducts. The contract was for 5500 units.

Figure 13.35 Grc permanent coffer units.

13.4 GENERAL APPLICATIONS

The following section is included as a collection of general applications of grc. It illustrates particularly the versatility of this material in a number of differing roles.

13.4.1 Foto Fair kiosks (Figure 13.36)

Shopping centre architecture is becoming more and more customer orientated. In response to this sophistication, Foto Fair International has redesigned its drive-up photo service kiosk. The free-standing units are usually located in the parking areas of major shopping centres in the States.

The new kiosk is a completely prefabricated thin shell module of grc. The previous unit was made of metal panels painted in a bright circus tent motif with a rotating sign on top. In contrast, the new kiosk is of a simple, contemporary design. It is entirely white except for the accent of yellow and red 'Foto Fair' logos which are an integral part of the mansard roof. The solid appearance of concrete suggests a responsible and professional approach which, it is hoped, will encourage customers.

Figure 13.36 Foto Fair kiosks.

The lightweight and good flexural strength of grc made prefabrication practical. Doors, windows, air conditioning, strip heating, counters, storage cabinets, lighting insulation—everything except the film supplies—are in place before a kiosk is shipped to a sales location.

At the shopping centre site, a concrete slab with a curb is prepared in advance and includes electrical conduit stubs. The 1900 kg module is then easily placed into position with a light crane. Finally two grc planter boxes are installed at each end of the slab. The planter boxes also conceal flood lamps for the night lighting. Figure 13.37 shows views during construction and assembly.

Figure 13.37 Views during construction and assembly.

The kiosk is produced in two sections using alkali resistant glassfibre and concrete. The 2.4 m × 1.2 m base section including walls, soffits and insulated sandwich floor, is formed in one piece. The 3.3 m × 1.65 m mansard roof section, also one piece, has a concealed outside duct for air conditioner intake and exhaust. Three-dimensional logos are also part of the roof form. The assembled structure stands 3.3 m high. Inside, four steel corner columns connect the base to the roof. The manufacturer ships assembled kiosks to another company who finishes the units and equips the interior, installs windows and doors and paints the logos and roof signs.

The cost of the grc kiosks is comparable to the older steel units, but they benefit from a more attractive appearance. The sculptured contours, recessed windows, the texture and integral logo were all design features possible with grc.

***Client:** Foto Fair International, Inc., Subsidiary of Progressive Industries Corp., Dayton, Ohio*

***Designer:** Richardson/Smith, Inc., Worthington, Ohio*

***Manufacturer:** Glas-Con, Inc., Bloomington, Minnesota*

13.4.2 Sun awnings, First and Lenora Buildings, Seattle, Washington (Figure 13.38)

Growing popularity of refitting old buildings is spurring numerous construction projects. One such project is the First and Lenora Building in Seattle, Washington. Previously a two-storey garage, the building was successfully adapted into a three-storey office building. The 68-year-old building is sturdy but with earthquake code requirements it could not support the precast concrete awnings desired by the architect and owner. Another alternative, plate aluminium, was troublesome aesthetically and also more expensive.

The awnings are a key design feature of the remodelling, giving depth and shadow to an otherwise flat exterior. Concrete was requested in order to blend with the cleaned-up existing portions of the concrete and brick exterior. The grc panels extend 1.5 m from the building face. They shade 2.1 m high windows from the afternoon sun, reducing the cooling load, and buffer the wind convection in the winter.

The thickness of the grc awning is only 12 mm. The underside has ribs spaced about 1.1 m apart, formed by applying grc over rigid foam strips. The panels are approximately 5.7 m wide and 2.4 m high but weigh only 636 kg.

The 20 grc awning panels were cast in plywood forms by first spraying a grey oPc mix formulated with a coarse sand selected to match the exterior of the existing building. This was then backed with the structural grc formulation. After curing, the top surface was lightly sandblasted to bring out the concrete appearance. Since the underside of the awnings is also exposed to view, the sprayed mixture was trowelled smooth and finished with a spray applied stucco finish. After curing, this surface was painted white to match the stucco trim around the windows.

The upper awning sections were located between the brick columns. However, the columns varied in dimension. To compensate for this variation the forms were fitted with blanks which allowed the awning panels to be made longer or shorter. To support the upper awning section steel angles were bolted through the spandrel beams and anchored to the grc inserts. Steel pipes mounted to the brick columns were used to brace the projecting section of the panel.

***Client and contractor:* Raper/Persham, Seattle, Washington**
***Architect:* R James, Dersham, Seattle, Washington**
***Manufacturer of panels:* Concrete Composites Incorporated, Seattle, Washington**

Figure 13.38 Sun awnings, First and Lenora Buildings, Seattle, Washington.

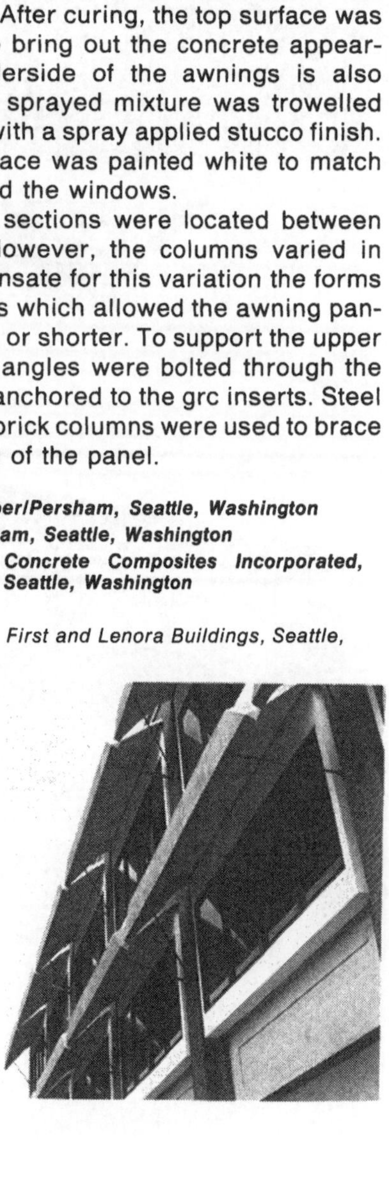

13.4.3 Promenade roof tiles

The grc promenade roof tile (Figure 13.39) has been designed as a protective cover for built-up bitumen roofs to guard against damage from pedestrian traffic and solar gain providing safe access to chimney and vented shafts. They are not designed to support vehicles or other superimposed loads and are unaffected by climate extremes normally experienced in the UK.

The tiles have a patterned top surface and are manufactured in grey and white. The nominal size is 300 mm × 300 mm × 15 mm. The edge is thickened to 15 mm over the 12 mm, and they weigh 2 kg. The tiles are easily cut with a power tool and abrasive disc and they are usually laid in hot bitumen.

Figure 13.39 Promenade roof tile.

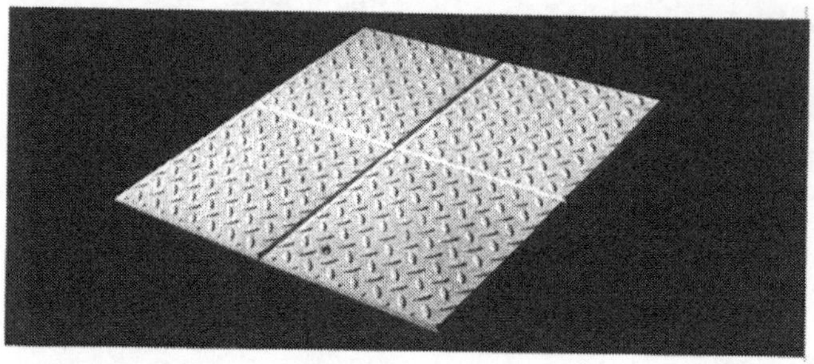

13.4.4 Fireplace surrounds

The ability of grc to simulate stone and brick has led to the production of a range of fireplace surrounds. An example in simulated stone is shown in Figure 13.40. For this, premix process was used on timber backing moulds, faced with a stone embossed plastic formwork liner. Other finishes have been produced to simulate brickwork traditional fireplaces with dressed stone surrounds, both in grc. Rubber moulds cast against stone or concrete masters have proved the potential of reproducing the correct texture by this method and the fire resistance of grc has enabled the fire grates to be sited directly onto the grc hearths.

Figure 13.40 Caravan fireplace surround.

Figure 13.41 Range of street furniture.

13.4.5 Street furniture

The mouldability and dense durable weathering of correctly manufactured grc makes it ideal for street furniture (Figure 13.41). A number of manufacturers have therefore developed products in both spray-up and premix grc.

The lightness of grc street furniture makes handling, transport and installation much easier and cheaper than the concrete equivalents. A number of textures and finishes have been tried on these products and the shiny, impermeable finishes have proved inconsistent in bonding to the surface of the grc. However, when these glossy surfaces have bonded, they offer a range of bright finishes that compares with plastic materials but has the strength and impact properties of concrete.

A notable fault of some street furniture has been observed when edges have been sharp and excessively thin. These edges can easily chip and break with heavy impact loads due to the lack of fibre penetration. The rim of such products should be kept to a 20 mm minimum thickness, thinning out progressively away from the edge.

13.4.6 Window frames

The durable nature of grc makes it an ideal replacement for wood manufactured window frames. Correctly formulated grc has particularly good weathering properties and maintenance commitments are therefore reduced.

Before considering grc as a simple material substitute it should be appreciated that, like all materials, its properties must be taken into account when examining uses and suitable design. It is not practical, for example, to try and simulate fine detailed sections on window frames which will be roughly handled on building sites. Also, window frames must be available in a vast range of sizes. It is, therefore, only sensible to consider a section that is strong when made in grc and can be cut and joined after curing in order to manufacture the size of window required.

The section shown in Figure 13.42 has proved successful for window construction—particularly by providing ideal jointing. The method of jointing is shown in Figure 13.43. It involves running the top and bottom members through and cutting off the required lengths with square cuts (a). The vertical members are cut to the section profile by the profile grinder (c). This is a mild steel arbor machined to match the window section. It can profile each end of the vertical members to produce a vertical section that, on assembly, has racking strength. The surface is then coated with diamond paste which is bonded to the arbor with resin.

The assembly is held together by rods located behind the vertical members and passing through

Figure 13.42 Practical grc window section.

Figure 13.43 Construction details of a grc window frame.

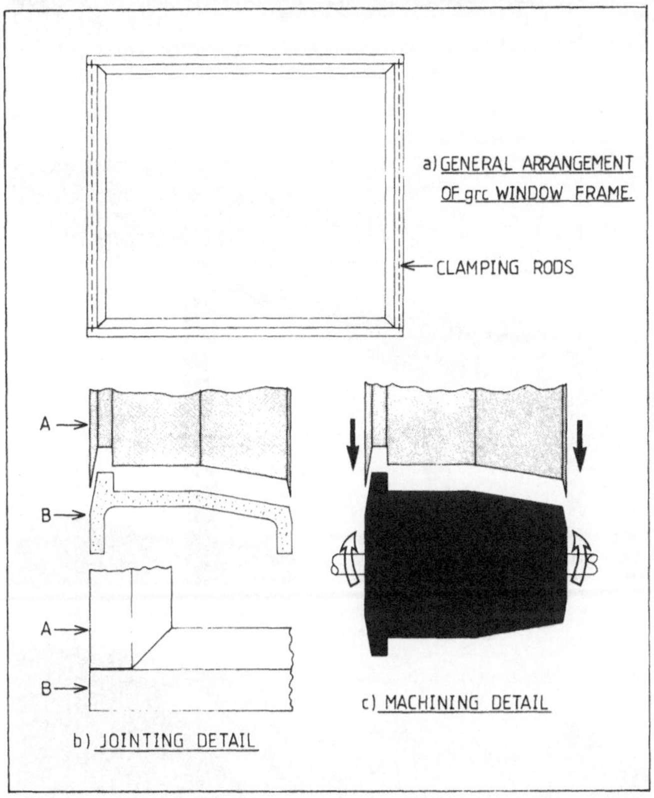

holes in the horizontal members. They are suitably clamped with nuts and washers. The joint detail is shown in (b). It may be finished with a mastic to provide a weather seal. The intention is that this frame will take either double-glazed units or aluminium frames. The finish can be in natural white oPc or possibly an epoxy coating as shown in Figure 13.42.

13.4.7 Application of coated 'E' glassfibre

Most recent developments of grc have centred on specially formulated alkali resistant glassfibre. However, a Dutch company Forton is now marketing a system that uses a coated 'E' glassfibre combined with a polymer admixture addition to the matrix. This system has been used on a number of contracts and the following illustrations show the potential of the system in cladding and housing applications:

- The Mondial Tiel furniture showroom is clad with 2.5 m × 2.0 m panels in white polymer modified 'E' glassfibre grc. The cladding is 12 mm thick and the eaves panels 10 mm thick. The eaves panels are also 2.5 m × 2.0 m (Figure 13.44).
- The panels on the Rotterdam abattoir are of sandwich construction 4.5 × 2.5 m and 160 mm thick, 140 mm being PIR foam (Figure 13.45).
- The houses in Figure 13.46 are both 6 × 9 m; the first of medium cost, the second low cost. All components including the roof panels in both houses are produced in polymer modified 'E' glass grc.

Figure 13.44 Mondial Tiel building.

Figure 13.45 Abattoir in Rotterdam.

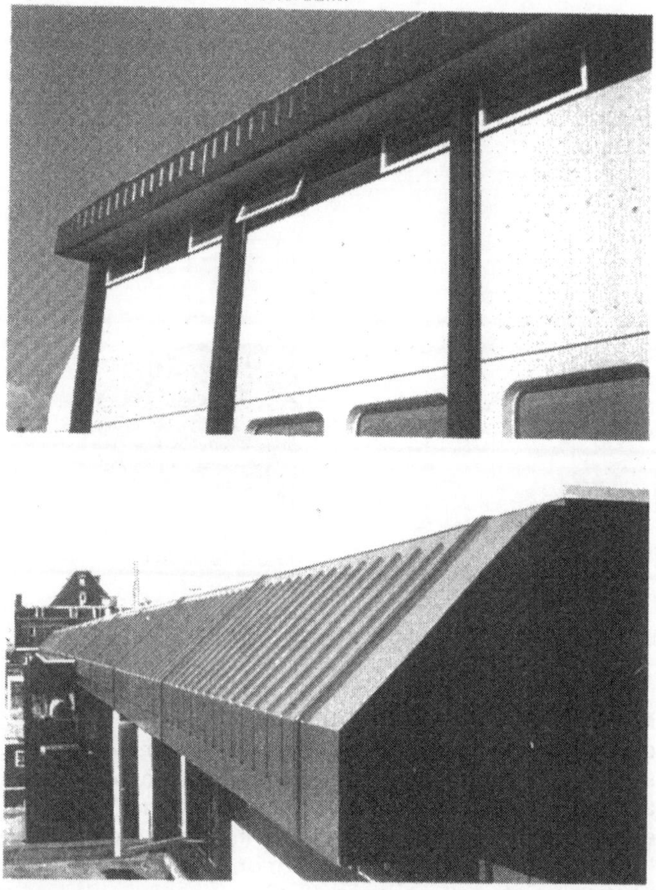

***Figure 13.46** Examples of houses with all components, including roof panels in polymer modified 'E' glass grc.*
(i) Medium cost shown above.
(ii) Low cost shown below.

14

The future

Glassfibre reinforced cement has now become established as a material with numerous applications mainly due to the development and marketing by Pilkington Brothers Ltd of Cem-FIL AR glassfibre. One consequence has been the development of other fibres and systems with the aim of capturing some of this market. These fall into two distinct categories:

(a) Fibres claiming to offer resistance to the alkaline environment formed by hydrating cement paste

(b) Systems using coated 'E' glass together with polymer solutions acting as matrix modifiers.

As new fibres become available that claim to be alkali resistant clearly some means of assessing and comparing this property is needed. The method of test must be a fair, impartial assessment based on the conditions that exist in hydrating cement paste. This prompts the question that, as yet, is not suitably answered—what do we mean by an alkali resistant fibre? The industry requires some means of categorizing alkali resistance in order to distinguish between these new fibres.

The second category mentioned above refers to systems using a combination process. The fibres are coated with a protective polymer binder and are supplemented by a polymer solution which is dispersed in the matrix. This system can utilize 'E' glassfibre which offsets the extra cost of the polymer dispersion in the matrix by the lower outlay on the 'E' glassfibre relative to alkali resistant grades.

Interest has also been shown in fibre combinations using fibres other than glass in order to improve the modulus of the composite in the safe working load and the load at failure. This work requires further verification.

Obviously the most topical and perhaps most commercially rewarding development focuses on replacing asbestos by alternative fibres. Products have recently come on to the market employing glassfibre reinforcement but difficulties exist in processing glassfibres for mass production using asbestos cement production machinery. The 'wet mat' properties of asbestos cement which enable large volume outputs to be achieved do not exist with normal grc mixes. Supplementary fibres have to be included.

Production equipment has undergone steady development and hopefully will continue particularly towards improvements in chopper gun design. This one item in the production process leads to more down-time than all the other machinery. Although the chopper gun is effective when working properly, it is readily found to slip and produce long fibres or to block. Setting-up after stripping down for maintenance is something of a 'hit and miss' process. Hopefully a new chopper gun will be developed that employs a more maintainable and constant performance.

As grc has become available it has presented itself as an alternative material to many others in common use. This had led to the situation that products can now be manufactured in grc providing direct competition to the conventional materials that are currently used. Consequently, where British Standards are in existence covering these products, they have to be extended to include grc. This requires care as the Standard must consider fairly the properties of grc.

Naturally, as new products are developed, new manufacturing processes will be developed to produce these products. Fibre contents will not necessarily be as high as 5% which was considered the norm in the past. For some applications, however, such as cladding, this will remain the recognized addition rate. Indeed, in the battle to replace asbestos, fibre contents will be much dependent on the properties required from the product. This may well lead to new Standards being evolved that take note of the distinctive properties offered by grc composites.

The grc industry has a manufacturers association, The Glassfibre Reinforced Cement Association, that is monitoring the situation with regard to new Standards and product specifications. This association is actively involved in drawing up Specifications and British Standards that are relevant to grc as a material and in product form.

The future for grc is good. Over the past 10 to 15 years recent developments have demonstrated its viability. It is increasingly fulfilling a distinct role as a material in its own right that must be assessed in terms of its documented properties.

15

Typical properties

Grc is not one material but a whole range of composites each one having its own properties dependent on the matrix used, the method of manufacture, age, environment and fibre content. Initial development and assessment of grc was carried out on neat cement paste matrices and therefore the most extensive property and durability data exist for these types of material. The results given in Table 15.1 are indicative of well-made grc with a neat cement paste matrix.

Present practice has favoured the inclusion of sand to the matrix to reduce drying shrinkage. The addition level is usually around 25% by weight of dry solids to the matrix. Admixtures may also be included as dispersing agents or to increase the strain capacity. This again produces a further set of material properties.

Such a variety of formulations make it difficult to be completely general about the properties of grc. Nevertheless, since most manufacturing processes use a few standard types of formulation, it is possible and practicable to consider these as a datum, relating any modification of the matrix and the subsequent property change to these established property values.

Manufacturing processes can also bring about property values particular to that process but consideration here is given to the current volume processes now in common use, namely spray-up, spray-up dewatered material and premix grc, with an additional mention of 'E' glass formulations having a polymer admixture.

15.1 GENERAL MATERIAL BEHAVIOUR AND AGEING CHARACTERISTICS

The general stress-strain curve of grc is shown in Figure 15.1. It is characterized by three regions. The extent and the detectability of these regions, however, may vary with the formulation and curing conditions as well as the fibre orientation.

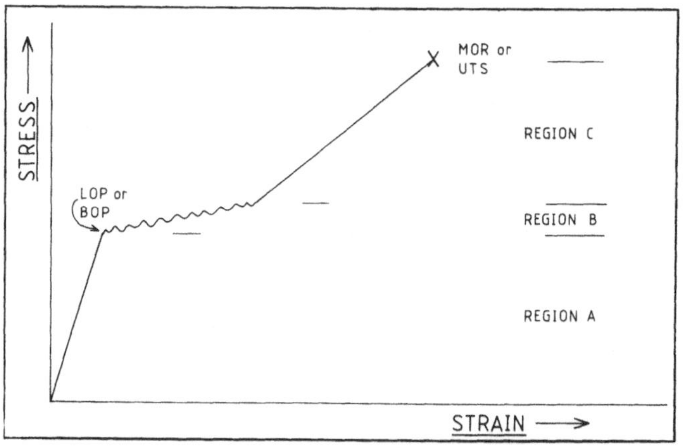

Figure 15.1 Idealized stress-strain behaviour of grc.

The first region A is a linear stress-strain plot where the matrix and fibres act together as a pseudo-elastic composite; the stiffness and stress distribution being a function of the Law of Mixtures. It should be noted that the matrix has a stiffness ≧ 3 times that of glassfibre. Thus:

$E_c = E_f V_f + E_m V_m$ and

$\sigma_c = \sigma_f V_f + \sigma_m V_m$

where:

E = Young's modulus
V = Volume fraction

Table 15.1 Strength properties of spray dewatered oPc/grc at various ages (5% glassfibre) BRE data.

	Total range for air and water storage conditions at 28 days	1 year			5 years			10 years		
		Air*	Water†	Weather‡	Air*	Water†	Weather‡	Air*	Water†	Weather‡
Bending strength										
MOR (N/mm²)	35–50	35–40	22–25	30–36	30–35	21–25	21–23	31–39	17–18	15–19
LOP (N/mm²)	14–17	9–13	16–19	14–17	10–12	16–19	15–18	14–16	16–17	13–16
Tensile strength										
UTS (N/mm²)	14–17	14–16	9–12	11–14	13–15	9–12	7–8	11–15	6–8	7–8
BOP (N/mm²)	9–10	7–8	9–11	9–10	7–8	7–9	7–8	9–10	6–8	6–8
Young's modulus (kN/mm²)	20–25	20–25	28–34	20–25	20–25	28–34	25–32	25–33	25–31	27–30
Impact strength 120D (Nmm/mm²)	17–31	18–25	8–10	13–16	18–21	4–6	4–7	15–22	2–3	2–6

*at 40% RH and 20 °C †at 18–20 °C ‡natural weathering

σ = stress
c = composite
f = fibre
m = matrix

Region A is not perfectly linear due to micro cracking which occurs normally with all cementitious materials. As the load is increased to a value approximating to the matrix failure stress, cracks propagate. The fibres bridging these cracks immediately carry the loads and transfer them back into the matrix, each side of the crack. As the loading is increased and the stress reaches failure level for the matrix further cracking occurs. This region B continues until the composite is traversed by a series of fine multiple cracks bridged by fibres at a spacing governed by the matrix strength, the bond strength of the fibres in the matrix and the fibre content. The region B is that of high energy absorption and considerable strain capacity. Once the composite has been crossed throughout its length by a grid of multiple cracks a critical crack, presumably traversed by fibres with either low bond strength or low fibre concentration or both, is the site of subsequent failure. As the load is further increased, the crack opens up until finally the two sections part (region C).

This failure pattern is the case in theory and, as will be shown later, may not be seen so readily in practice. The change in region A to region B is referred to as the LOP (limit of proportionality) in bending tests or the BOP (bend-over point) in tensile tests. The maximum stress level of region C is known as the MOR (modulus of rupture) or UTS (ultimate tensile strength) in bending and tensile tests respectively.

Present durability data indicate that, in dry storage conditions, modulus of rupture, tensile strength and strain to failure remain high but there is a fall in LOP and BOP levels. On the other hand, once stored in wet conditions (for example, 1 year in water, or 5 years in natural weathering), grc becomes essentially brittle (Figure 15.2); the tensile stress-strain curve terminates just beyond the bend-over point. The MOR reduces considerably and nearly meets the LOP value. However, LOP and Young's modulus both increase to some extent in these conditions (Figure 15.3).

In dry air storage, changes in strength occur slowly and are relatively small. With water immersion there is

Figure 15.2 Strain-to-failure in tension for grc at various ages in dry air, natural weathering and water storage.

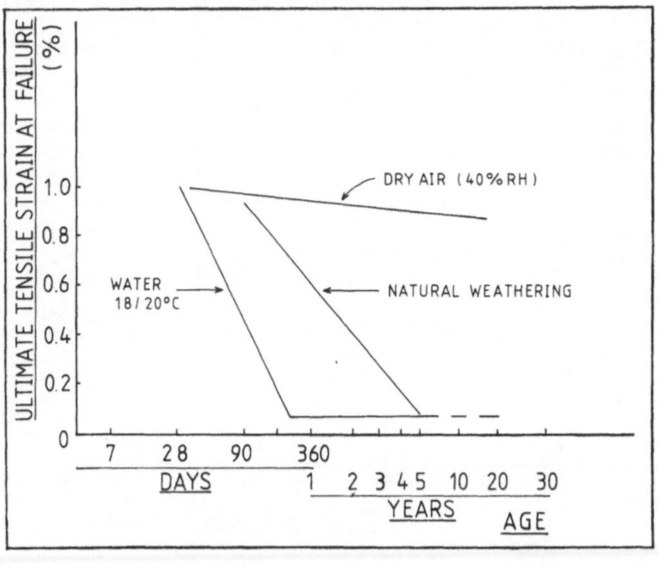

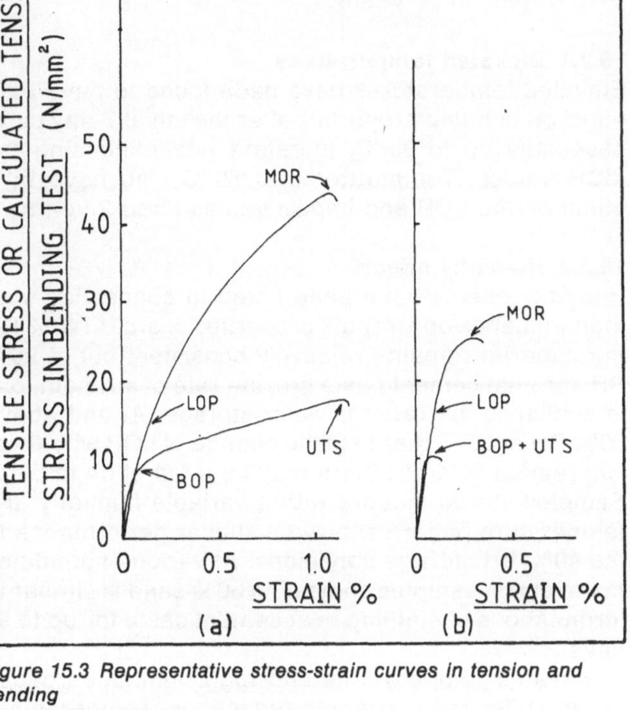

Figure 15.3 Representative stress-strain curves in tension and bending
(a) After 28 days in water at 18–20 °C
(b) After 5 years in water at 18–20 °C.

a rapid fall in strength which then remains almost constant over 1 to 5 years. Temperate natural weathering conditions result in a loss in strength over a 5-year period.

The factors that lead to the change with time in composite properties are not fully understood and no theoretical method for determining the shape of the durability curve exists at present.

Although careful attempts have been made to measure the UTS of specimens aged under dry, wet and natural weather conditions, it is universally recognized that it is very difficult to make such measurements on brittle materials. Any departure from 'perfect' experimental conditions (such as misalignment and stress concentrations at grips) will inevitably lead to an underestimate of the UTS. This becomes particularly important with grc after embrittlement in wet or weather storage.

15.2 MECHANICAL PROPERTIES IN AIR STORAGE CONDITIONS

The available data are concerned with material stored in controlled 40% RH conditions at temperatures of 18 °C for the first 2 years and 20 °C for the remainder of the time. Four point-bending tests were carried out on specimens of 150 × 50 × 9 mm over a span of 135 mm. The tensile tests used simple strip specimens of the same size and the impact tests were carried out on the broken halves of the flexural specimens.

Under these dry storage conditions there is a small fall-off in strength properties in the early years but the material retains its tough pseudo-ductile behaviour at 5 years. On the basis of present trends best estimates of average strength properties at 20 years are MOR 31–39 N/mm^2, UTS 11–15 N/mm^2, impact 15–22 N/mm^2. Over the first 1 to 2 years in these conditions a fall in

LOP is experienced; this seems to rise and stabilize at 20 years (estimate) to 14–16 N/mm^2; BOP would be 9–10 N/mm^2 at 20 years.

15.2.1 Elevated temperatures

Elevated temperatures have been found to have little effect on bending properties after the initial 7-day cure, especially up to 35 °C, therefore not influencing the MOR values. Temperatures to 50 °C also have little effect on the LOP and impact values up to 2 years.

15.2.2 Humidity effects

Humidity effects have been noted in connection with their influence on strength properties of grc. At 40% RH the material remains relatively consistent but at 90% RH, for neat cement paste grc, the rate of strength loss is similar to the rate for water storage. At and above 60% RH at 20 °C there is little change of LOP with time, whereas at 40% RH there may be a fall of up to 50%. Samples stored indoors with a variable humidity and temperature pattern showed a similar performance to the 40% RH storage conditions. At various humidities the effect on samples containing 30% sand is similar to formulations containing neat cement paste for up to 90 days.

Elevated temperatures and high humidities cause grc to suffer rapid strength reductions. Temperatures of 50 °C and humidity levels of 60–70% RH can lead to MOR values dropping to 16–19 N/mm^2 in 30 days but thereafter these values are maintained for the next 30 days.

15.3 MECHANICAL PROPERTIES IN WATER STORAGE CONDITIONS

Results of tests to determine MOR, LOP, UTS and BOP and Izod impact strength after 5 and 10 years immersion in water at 20 °C (Table 15.1) together with strain to failure data (Figure 15.3) show embrittlement and strength loss over the first year with subsequent stabilization at near constant values. The samples exhibited, however, a small gain in LOP when stored in water over this period.

The stress-strain curve obtained with simple strip-type tensile tests on wet aged material indicates almost entirely brittle behaviour with only a very small bend-over region. The MOR also remained somewhat above the LOP (Figure 15.3), and MOR and impact strength values were about two to three times cement paste values.

15.3.1 Temperature effects

Present results show that increases in the temperature of water storage cause more rapid and more severe losses in strength of grc. The magnitude and extent of these effects is evident at 50 °C when the MOR and LOP is seen to merge completely at 3 months. This indicates a more complete embrittlement than at the lower water storage temperatures.

15.4 MECHANICAL PROPERTIES IN NATURAL WEATHERING CONDITIONS

Over 10 years of available exposure data (air temperatures −10 °C to 31 °C and 30% RH → 100% RH, annual rainfall 590 mm) results indicate that for up to 5 years the strength properties fall more slowly than in water storage conditions. They have, however, reached approximately the same levels at the end of that period and appear to be still falling.

With limited data available and the longer time scales involved in these strength changes it is very much more difficult to make valid predictions of properties at 20 years than in the case of water storage. The LOP values remain stable in temperate weather conditions and the BOP becomes virtually coincident with the measured UTS after about 2 years. Weathering data are also being accumulated on a range of sites in widely differing climatic conditions. After preliminary results at 1 year, two situations emerge as giving noticeably different results from tests carried out at the Building Research Establishment. In the hot, wet conditions of Singapore there is a very rapid drop in strength properties over the first 6 months, with an indication of levelling out at 1 year. In Toronto, samples subjected to freeze-thaw conditions show signs of a fall in LOP level.

Data published by Pilkington Brothers for the effect of natural weathering on Cem–FIL composites are shown in Figures 15.4, 15.5 and 15.6.

15.5 YOUNG'S MODULUS VALUES

The Young's modulus values of 10-year old grc for the three environments used are listed in Table 15.1. These values have been estimated from the tensile stress-strain curves. The stiffness of grc is dependent on the stiffness of the matrix. It is clear from the results that the values change little between 1 and 10 years, but are somewhat higher than the start levels.

15.6 CREEP, STRESS RUPTURE AND FATIGUE BEHAVIOUR

Measurements of creep in bending and direct tension have been made for indoor and water storage conditions. In direct tension creep strains are very much smaller than expansion-contraction strains due to humidity.

Bending tests show that at all stress levels creep

Figure 15.4 MOR and LOP in UK weather.

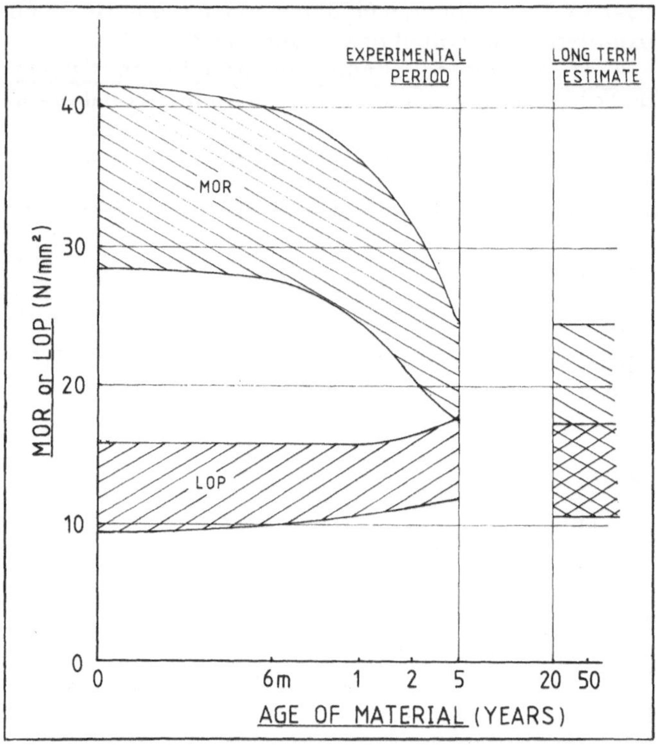

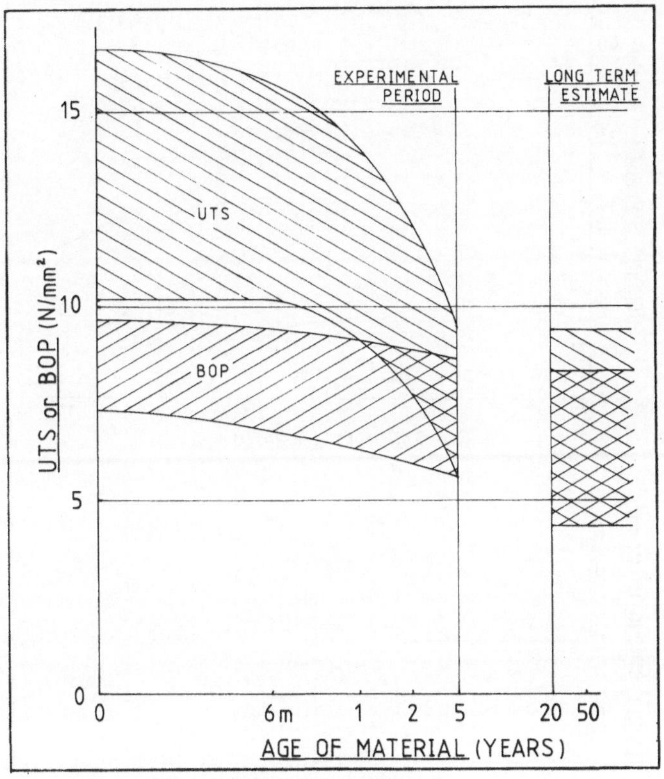

Figure 15.5 UTS and BOP in UK weather.

Figure 15.6 Izod impact strength in UK weather.

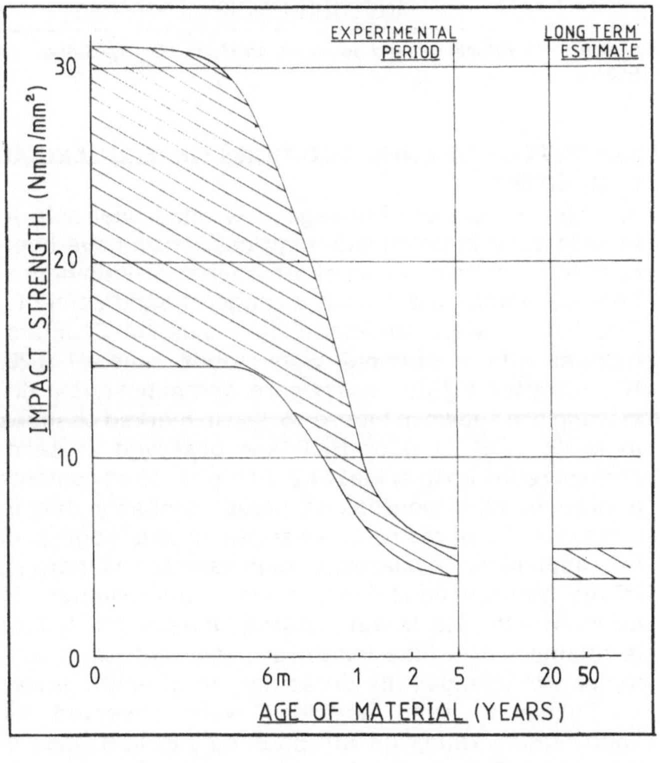

behaviour of composites is qualitatively similar to that of cement. At stresses less than the LOP the creep is in fact quantitatively the same as that of the matrix material alone when made under the same conditions and loaded at the same time. Figures 15.7 and 15.8 show the ratio of creep strain to initial strain for applied bending stresses below the LOP level in indoor and water storage respectively and at various loading times. Most data are for a cement paste matrix with a water/cement ratio of 0.3 but also included in Figure

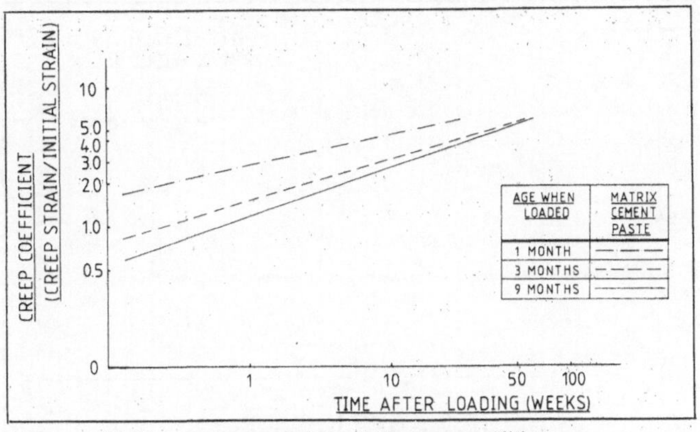

Figure 15.7 Flexural creep of grc in indoor conditions.

Figure 15.8 Flexural creep of grc in water.

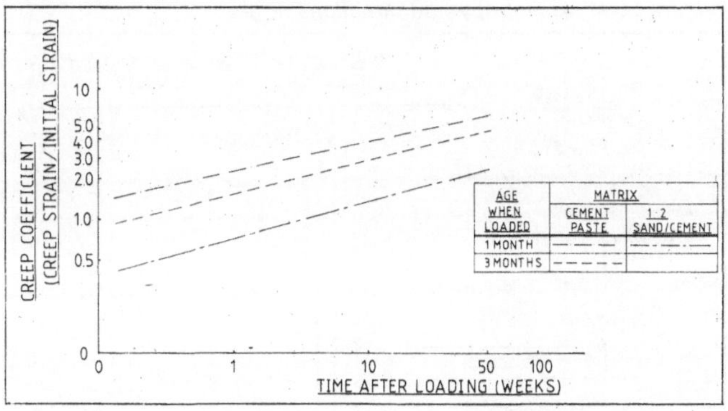

15.8 are results for a 1:2 sand:cement mortar matrix with the same water/cement ratio showing a reduction in creep. This reduction may not however occur if the water/cement ratio is allowed to rise when adding the sand.

Stress rupture experiments under constant bending load in water and indoor conditions have shown no failure below an applied stress level of 19.5 N/mm^2 in 2 years. In temperate weathering conditions there had been two stress rupture failures at 18 N/mm^2 at 18 months out of 12 samples at that stress level.

Fatigue tests have been carried out in bending and direct tension. Bending results show that lives of greater than 10^5 cycles are obtained in wet and dry conditions at the LOP stress (Figure 15.9) with lives in excess of 10^6 cycles at 10 N/mm^2. Direct tension tests show a life of about 2×10^4 cycles at the BOP and a life of 10^6 cycles or above at 4.5 N/mm^2 in both prolonged

Figure 15.9 Flexural fatigue of grc in air and water.

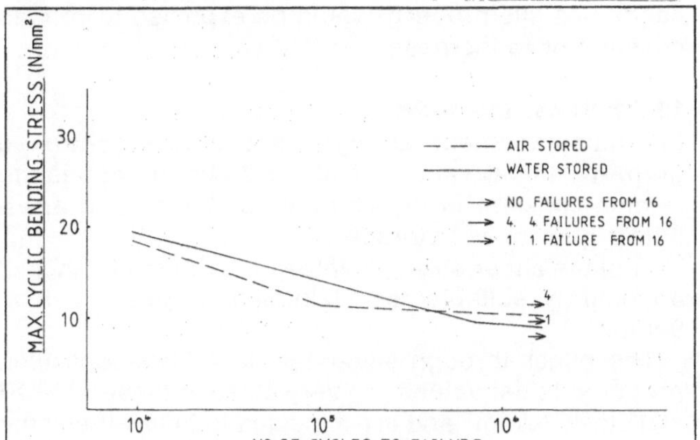

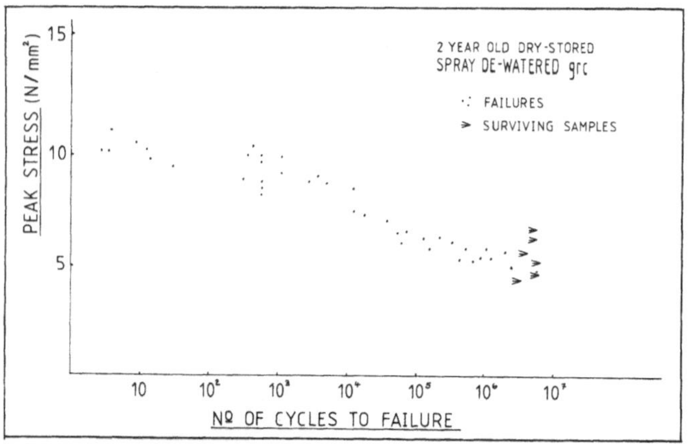

Figure 15.10 Tensile fatigue of grc in air.

Figure 15.11 Anisotropy and directions in grc.

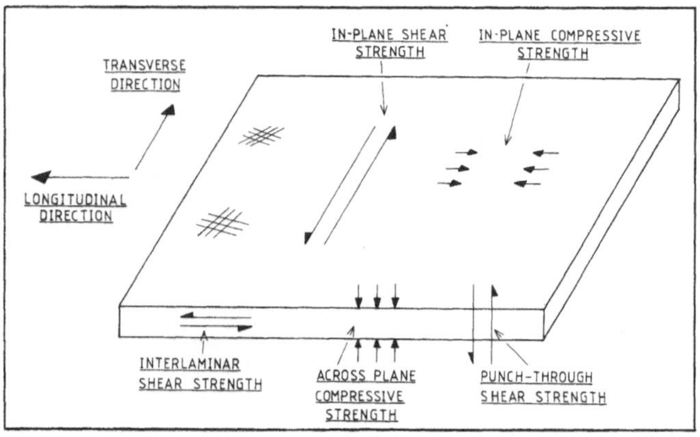

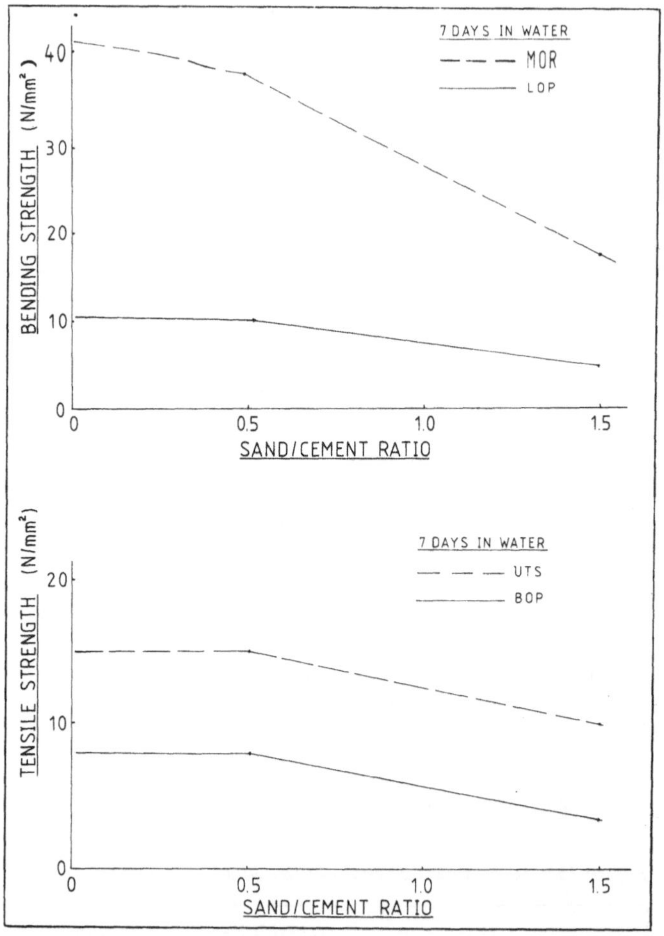

Figure 15.12 Effects of sand additions on MOR, LOP, UTS and BOP.

wet and dry storage conditions. Figure 15.10 shows results for 2-year-old indoor stored material.

15.7 COMPRESSIVE AND SHEAR STRENGTHS

These properties are considerably dependent on the direction of test and are illustrated in Figure 15.11 with reference to the terminology used.

15.7.1 Compressive strengths

The across-plane compressive strength of grc is equal to that of cement paste at about 80–100 N/mm^2. There is no indication of a fall with time over 6 months dry storage and in wet conditions this property would be expected to remain constant or increase.

In-plane compressive strength is somewhat lower at around 50–60 N/mm^2 and again there is no indication of a fall with time over 6 months dry storage. In wet conditions this property would be expected to remain constant or to increase.

15.7.2 Shear strengths

Inter-laminar shear strength is a matrix controlled property. Initial values of about 2 N/mm^2 appear to remain constant in dry storage conditions and are a little higher in wet storage.

In-plane shear strength values are limited by lack of an entirely suitable test. Minimum values are 8 to 9 N/mm^2.

The punch-through shear test is a fibre-controlled property. Initial values are very close to those of MOR at 35 to 40 N/mm^2 and are expected to follow the same durability pattern as MOR in different environments.

15.8 EFFECT OF SAND ADDITIONS ON MECHANICAL PROPERTIES

In order to control shrinkage it is often desirable to include sand in composite matrices. Tests have been carried out with a series of grc boards containing 5% fibre by weight and having a range of sand contents. The boards were all made by Pilkington Brothers, Lathom with a nominal water:solids ratio of 0.26. Results after a 7-day water cure and subsequent dry storage are shown in Figure 15.12. No marked changes in MOR, LOP, UTS or BOP are observed at sand: cement ratios up to at least 1:2. At higher sand contents a drop in all properties is noted, probably due to increases in porosity or changes in the degree of hydration in these boards. At high sand levels there is, in dry storage conditions, a small improvement in strength with time. In water storage the strain to failure is retained for a little longer at these high sand contents. No changes in durability or strength levels ascribable to sand additions were observed for sand:cement ratios up to about 1:2 (up to 1 year in water and air).

15.9 THERMAL EXPANSION COEFFICIENTS, THERMAL CONDUCTIVITY AND GAS AND WATER VAPOUR PERMEABILITY OF GRC

Thermal conductivity of grc depends more or less linearly on density of the composite and varies between 0.5 W/m °C at a bulk density of 1700 kg/m^3 and 1.1 to 1.3 W/m °C at 2200 kg/m^3.

Thermal expansion coefficient of dried grc lies between 7 and 12 $\times$ 10^{-6} per °C but for damp material it may be as high as 36×10^{-6} per °C under certain conditions of atmospheric humidity.

Material age, storage conditions and water/cement ratio effect the various permeability values of grc.

It should be noted that:

(a) Air permeance increases with storage in dry (40% RH) conditions. A value of 3 metric perms was recorded at 1 year with some further increase probably indicated by the trend of results. In high humidity (90% RH) and wet storage conditions permeance was < 0.15 metric perms.

(b) Water vapour permeance values were 7 metric perms for a board made at 0.35 water/cement ratio, and fell slowly over 1 year. For a board made at 0.25 water/cement ratio the water vapour permeance was ≧ 3 metric perms.

(c) Grc passes the ISO permeability test at an age of 28 days.

15.10 DENSITY

The density of grc varies markedly with the fabrication technique and the matrix employed. Normally the range conforms to 1700–2000 kg/m^3 for spray-up material and 2000–2200 kg/m^3 for spray-dewatered material.

15.11 FIRE PERFORMANCE

In assessing the fire performance of grc, it must be remembered that grc is a family of materials and not a single material. Consequently it has fire properties directly dependent on the matrix formulation and fabrication method.

The UK assess the fire performance of a structure to BS 476: Part 8: 1972. This test has 3 criteria:

- Stability—the structures must not collapse under test
- Integrity—the structure must not allow flames to penetrate
- Insulation—the protected side must not exceed 140 °C above the initial temperature at the start of the test.

Single skin grc will not satisfy the insulation criterion and indeed for a formulation of cement pfa an air entraining agent is required to guarantee integrity. This latter formulation is only recommended for internal applications. To enable a 1-hour fire rating to be achieved using single skin grc, a suitable insulant such as a vermiculite or perlite surface skin is required. This then satisfies all three criteria of the BS test.

Sandwich panels encasing an insulating core can easily achieve up to 4-hour ratings, but obviously are expensive to produce. Figure 15.13 indicates the fire resistance of particular formulations and Table 15.2 gives the construction performance to BS476: Part 8.

It should be noted that BS 476: Part 8 refers to the whole construction under test, including joints and fixings.

15.12 MOISTURE MOVEMENT

Work carried out to determine the effects of moisture movement on grc, especially spray-up material, indicate that sand additions of 25% reduce the ultimate shrinkage to 0.15% for dewatered material and up to 0.2% for non-dewatered grc. However, it would not be expected that, for example, a vertical cladding panel would in a temperate climate experience a movement of 0.15 mm/m. Again, the exterior of the panel would affect the performance if coatings or aggregate finishes were present. Thus 0.075 mm/m would be a more likely response.

Higher sand contents reduce moisture movement but at the cost of reducing the strength properties of grc, due to the higher water demand required to make the matrix workable at the mixer.

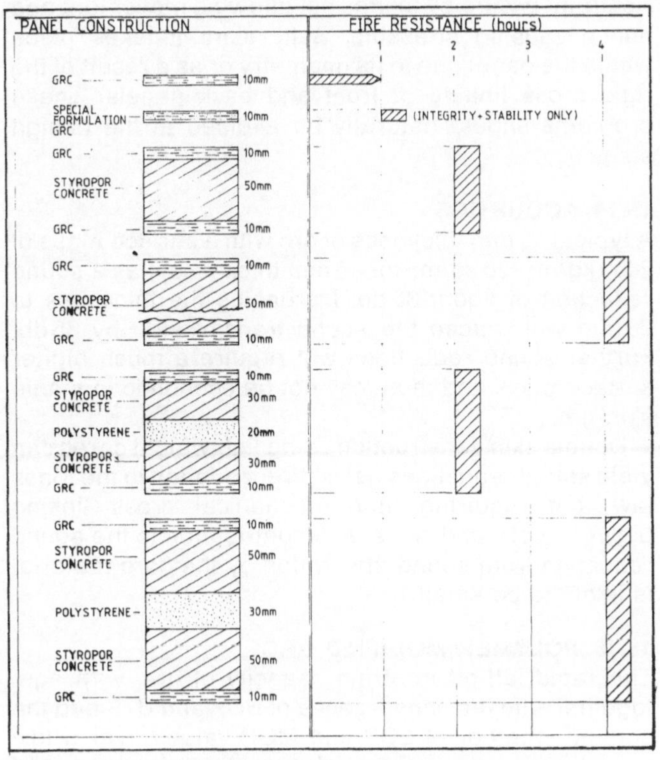

Figure 15.13 Fire resistance performance of grc panel construction.

Table 15.2 Construction performance to BS 476: Part 8.

Test	Title	Performance
BS 476: Part 4, 1970 (ISO R 1182)	Non-combustibility test for materials	Non-combustible
BS 476: Part 5, 1968	Ignitability test for materials	Class P (not easily ignitable)
BS 476: Part 6, 1968	Fire propagation test for materials	Typical test values I:0.0 (i_1):0.0
BS 476: Part 7, 1971	Spread of flame test for materials	Class 1 (very low surface spread of flame)

15.13 TEMPERATURE STRESSES

A restrained panel, such as a cladding panel with an insulating core, experiences a stress differential across its cross section due to a temperature gradient through the unit. The outer skin could be about ±20 °C different to the inner core. The strain produced is:

$$12 \times 10^{-6} \times 20 = 240 \times 10^{-6}$$

alternatively expressed as 240 µ strain.

Assuming uniform distribution front and back of the panel ($E = 20 \times 10^3$ N/mm^2):

$$\sigma \text{ thermal} = (\tfrac{1}{2})\, 240 \times 10^{-6} \times 20 \times 10^3$$
$$= 2.4\ \text{N/mm}^2$$

Restraint occurs by fixings not allowing movement and hence causing stressing. Self restraint takes place within the panel due to its geometry or as a result of the rigid cross linking of front and back panels. These problems should naturally be avoided at the design stage.

15.14 ACOUSTICS

A typical 10 mm thickness of grc with a surface mass of 2000 kg/m^3 (20 kg/m^2 for 10 mm thickness) has a sound reduction of about 30 db. Increasing the thickness to 20 mm will reduce the sound transmission by 35 db. Further sound reductions will require a much higher surface mass and thus may not be economic in single skin grc.

Double skin construction using lightweight cores can yield sound reductions not in direct relation to the Mass law, but requiring little mechanical cross linking between core and skins. Any contribution to the sound reduction from sound absorption by the core material is likely to be small.

15.15 POLYMER MODIFIED GRC

The rapid fall-off in strain capacity of grc, with age, together with the convergence of BOP and UTS and the coming together of LOP and MOR values over a few years' curing has prompted investigations to try and extend this pseudo-ductile region over a greater time scale.

One reason for this reduction in properties is that the composite structure of grc after manufacture leaves spaces of 3 microns between glassfibre strands. The cement matrix cannot occupy this space because the minimum size of its components is about 30 microns. As hydration progresses, however, hard crystals of mainly calcium hydroxide develop in these spaces producing a firm link between matrix and fibre reinforcement. This results in brittle stress-strain performance and low strain capacity and failure even when loads are below the design values.

Techniques have been developed to impregnate the fibre bundles with soft elastic polymers (mainly acrylics) which prevent the ingress into the fibre bundle of the hard calcium hydroxide crystals. This impregnation has also led to glass fibre protection from the alkali matrix, so that non-alkali resistant glass fibres have been used successfully. The level of polymer inclusion is around 15% by volume.

The properties obtained are as follows:

Young's modulus	6 kN/mm^2 after 28 days 10.5 kN/mm^2 after 1 year (both at 20 °C and 65% RH)
Drying shrinkage	1200 microstrain
Linear thermal expansion	12×10^{-6} °C
Water vapour diffusion resistance	$\mu = 550$
Water vapour permeability coefficient	$\sigma = 0.3 \times 10^{-12}$ s
Water absorption coefficient	10^{-3} kg/m^2 s$^{\frac{1}{2}}$
Water penetration coefficient	0.04×10^{-3} m/s$^{\frac{1}{2}}$
Density	1750 kg/m^3
Creep	Below BOP and LOP is as for matrix. After 1 year at 20 °C and 65% RH the creep at 8 N/mm^2 is 2.5%. The creep coefficient is 5.0.

Further properties are given in Table 15.3 and Figure 15.14. The latter shows the stress-strain plot for polymer modified grc under natural weathering conditions. This should be considered against the published figures of Cem–FIL grc having a tensile strain to failure of 1.0% at 28 days, 0.5% at 1 year and 0.05% at 5 years for natural weathering,

Table 15.3 Properties of polymer modified grc.

	28 Days	After 10 years: Dry 65% RH 20 °C	Under water	Natural weathering
Bending strength MOR (N/mm^2)	27	21	19	22
Limit of proportionality LOP (N/mm^2)	12	15	14	15
Tensile strength UTS (N/mm^2)	12	13	12	13
Bend-over point BOP (N/mm^2)	8	10	10	10
Elongation at rupture %	6	2.5	1.5	2.5
Interlamina shear (N/mm^2)	4	4	5	4
Charpy impact strength (Nmm/mm^2)	15	7	4	8

Figure 15.14 Tensile-strain curve for polymer modified grc ('E' glassfibre).

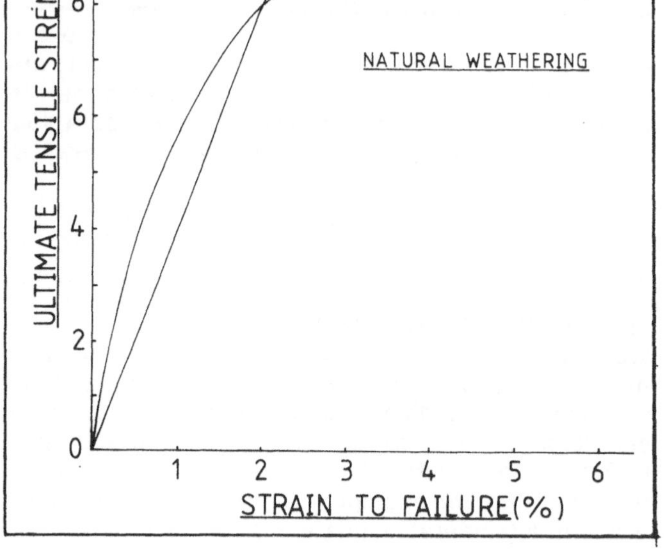